WHAT WOULD MACGYVER DO?

BRENDAN VAUGHAN is a senior editor at *Condé Nast Portfolio* magazine. He lives in Brooklyn with his son, Roan, his daughter, Dory, and his wife, Melissa, who knows a lot more about electric wiring than he does.

What Would MacGyver Do?

True Stories of Improvised Genius in Everyday Life

BRENDAN VAUGHAN

Neither MacGyver the TV series nor Paramount Pictures
authorized or had anything to do with the making of this book.
The stories within are based on real-life incidents of
MacGyver-esque ingenuity.

A PLUME BOOK

PLUME
Published by the Penguin Group
Penguin Group (USA) Inc., 375 Hudson Street, New York, New York 10014, U.S.A. •
Penguin Group (Canada), 90 Eglinton Avenue East, Suite 700, Toronto, Ontario, Canada
M4P 2Y3 (a division of Pearson Penguin Canada Inc.) • Penguin Books Ltd., 80 Strand,
London WC2R 0RL, England • Penguin Ireland, 25 St. Stephen's Green, Dublin 2,
Ireland (a division of Penguin Books Ltd.) • Penguin Group (Australia), 250 Camberwell
Road, Camberwell, Victoria 3124, Australia (a division of Pearson Australia Group Pty.
Ltd.) • Penguin Books India Pvt. Ltd., 11 Community Centre, Panchsheel Park, New
Delhi–110 017, India • Penguin Group (NZ), 67 Apollo Drive, Rosedale, North Shore
0632, New Zealand (a division of Pearson New Zealand Ltd.) • Penguin Books (South
Africa) (Pty.) Ltd., 24 Sturdee Avenue, Rosebank, Johannesburg 2196, South Africa

Penguin Books Ltd., Registered Offices: 80 Strand, London WC2R 0RL, England

Published by Plume, a member of Penguin Group (USA) Inc. Previously published in a
Hudson Street Press edition.

First Plume Printing, January 2008
10 9 8 7 6 5 4 3 2 1

Diagrams by Jim Kopp

Ⓟ REGISTERED TRADEMARK—MARCA REGISTRADA

The Library of Congress has catalogued the Hudson Street Press edition as follows:
Vaughan, Brendan.
    What would MacGyver do? : true stories of improvised genius in everyday life / Bren-
dan Vaughan.
        p. cm.
    Includes index.
    ISBN 1-59463-024-0 (hc.)
    ISBN 978-0-452-28929-1 (pbk.)
    1. Creative ability in technology—Anecdotes.   2. Inventions—Miscellanea.
    3. Resourcefulness—Folklore.   I. Title.
    T49.5.V38   2006
    600—dc22   2006023707

Printed in the United States of America

PUBLISHER'S NOTE
Neither *MacGyver* the TV series nor Paramount Pictures authorized or had anything to
do with the making of this book. The stories within are based on real-life incidents of
MacGyver-esque ingenuity.

For Melissa and Roan

Contents

What Would MacGyver Do?

Introduction

*"A paper clip can be a wondrous thing. More times than
I can remember, one of these has gotten me
out of a tight spot."* —MacGyver

IF YOU SOMEHOW BOUGHT THIS BOOK WITHOUT KNOWING WHO
MacGyver is—well, God bless you. That's quite a leap of faith.

As the rest of you are well aware, Angus MacGyver was
the greatest action-adventure television icon of the 1980s.
MacGyver, starring the mullet-headed Richard Dean Anderson,
debuted on ABC on September 29, 1985, and lasted seven
magical seasons. The title character was a Midwestern secret
agent who refused to carry a weapon, relying instead on his
wits—specifically, his ability to solve any problem using basic
scientific principles and whatever materials happened to be
lying around—to foil the enemy, usually in the nick of time.

Of course, this book would not exist if MacGyver were
merely a television program, even one produced by Henry
Winkler (who definitely has a knack for creating iconic TV
characters who are somehow cool and nerdy at the same
time). Long before the series finale aired on April 27, 1992,
MacGyver had begun to transcend *MacGyver*, entering the

vernacular as a verb: To "MacGyver" something is to fix it in a resourceful and improvisational way, e.g., "Our raft sprang a leak halfway down the river, but I MacGyvered a patch with my Swiss Army knife, a condom, and some tree sap, and we made it to the takeout." There's also a "MacGyverism," which can be either a) the act of "pulling a MacGyver," or b) the thingamajig—the makeshift invention—created in the process. And finally, there's the proper noun as honorific, bestowed upon especially creative handymen (and -women), as in, "Nice job, MacGyver."

What's equally remarkable, though, is MacGyver's staying power as a cultural reference. Almost fifteen years after the show ended, the character lives on, bubbling through our culture in strange and wonderful ways. On *The Simpsons*, Marge's spinster sister, Selma Bouvier Terwilliger Hutz McClure, is a huge fan. (She ritually treats herself to a smoke after *MacGyver*.) A couple years ago, McDonald's ran a TV ad in which a beleaguered mom stared at the camera and wondered, "What would MacGyver do?" Ice Cube, Big Punisher, and a handful of other rappers have mentioned Mac in their songs. *ReadyMade*, an excellent bimonthly magazine for people who like to make new stuff out of old stuff, has a regular section called "The MacGyver Challenge," in which readers compete to build something useful from a list of simple parts (a defunct radio, for example, or an old shipping pallet).

But two specific moments in MacGyver's afterlife tower above all others. One was in February 2006, when Mac scaled the Everest of American pop culture: He appeared in a commercial during the Super Bowl. It was the fourth quarter. In a perfectly executed gadget play, Antwaan Randle El of the Pittsburgh Steelers had just thrown a touchdown pass to his fellow wide receiver Hines Ward. Steelers 21, Seahawks 10. Game (basically) over. And there, all of the sudden, a few pounds heavier, a few steps slower, and without his mullet, was Richard Dean Anderson: MacGyver! He was tied to a chair in a warehouse; a time-bomb ticked nearby. Fortunately, he'd just used his MasterCard to purchase a tree-shaped air freshener, some nasal spray, a pair of tube socks, and a turkey baster, all of which came in handy as he MacGyvered his way to safety, mere seconds ahead of the blast. The tagline: "Little things that get you through the day: Priceless." An estimated ninety million Americans saw that ad, compared to the 13.1 million who watched the *MacGyver* finale in 1992. And despite the fact that Anderson hadn't played the character since *Trail to Doomsday*, a made-for-TV movie based on the show that aired in November of 1994, MasterCard saw no need to remind viewers who they were watching.

The other huge moment for MacGyver came in August 2003, when Ira Glass of NPR's *This American Life* (the cultural opposite of the Super Bowl) produced a terrific episode about real-life MacGyverisms. That episode, which you can still hear at thislife.org, inspired this book. In fact, Susan Burton's story was originally written for it, as was Chuck Klosterman's.

So MacGyver has entered the lexicon—as a noun, as a verb, as a symbol for craftiness, quick thinking, and a good, clean brand of improvisational genius. MacGyver has endured. As an icon, a legend, he is here to stay. Why?

My theory is that it's because we live in an age of declining know-how. There's a genuine crisis of competence: Nobody knows how to *do* anything anymore. Consider this statistic: In 1950, 46 percent of America's gross domestic product came from the "service" industries, i.e., businesses that perform services rather than make goods. By 2004, that number was over 60 percent. Not so long ago, a man was expected to know how to change his oil, unclog a drain, and maybe even do some basic carpentry. Now there's a thriving cottage industry of handymen who will come to your house and assemble the cabinet you just bought at Ikea. (You can find them on Craigslist.) In short, we've become a nation of specialists. We're all good at the one thing we're paid to do, but clueless about most everything else; this makes a supremely competent guy like Mac-Gyver even more exotic (and admirable). And this appeal is universal. No matter where you live or what you do for a living or how old you are, who can deny the basic satisfaction of solving the problem and saving the day? Bottom line: We'd all love to be MacGyver, if only for one shining moment.

The forty-two people who contributed to this book pulled it off.

It would be logical to assume that, as the editor of *What Would MacGyver Do?*, I've got a little MacGyver in me. I don't.

The truth is, MacGyver would be ashamed of me. I'm intimidated to the point of paralysis by mechanical things. I've never used a paper clip for anything other than clipping paper. And I have no real gift for improvisation. Like most of us, I always think of the perfect solution (or the perfect comeback) too late, after the opportunity to seize the moment has long since passed.

What I do have, however, is a deep well of awe and admiration for those who possess the MacGyver gene. And that's why I was so eager to take on this project. I was fourteen when *MacGyver* premiered, and watching those early episodes, I remember being conscious that there were people in the real world whose minds were MacGyveresque—and that I wasn't one of them. My hero in high school was my friend Doug Brazy, who completely restored a '65 Mustang at the age of sixteen and was always responsible for designing and building our class's homecoming float. I begged Doug to contribute a story to this book, but he was too busy working for the National Transportation Safety Board. (They're the people who, among other heroic duties, salvage the black boxes from plane crashes and try to figure out what went wrong.) As I sifted through the stories that make up this collection, my sense of awe for people like Angus MacGyver and Doug Brazy only grew—and then turned to full-on jealousy.

Before I get into which stories I selected (and why), a word on how I collected the submissions. First, I built a Web site: WhatWouldMacGyverDo.com. The site served as the hub of the project; it included a FAQ section explaining the concept, a simple mechanism for submitting stories, and a few sample MacGyverisms. Then I invited everyone—friends, acquaintances, friends of acquaintances, acquaintances of friends, writing students, science teachers, science students, the readers of *ReadyMade* magazine, etc.—to contribute. And contribute they did. Altogether, I received hundreds of submissions. Some weren't more than a sentence or two. ("Once I used to duct tape to fix my brother's glasses.") Others were great stories but lacked a true MacGyver moment. Others had solid MacGyver moments but lacked a strong story. And of course there were some that had no redeeming qualities whatsoever.

Now I'd like to introduce you to some of the authors whose stories did make the cut. I'm talking about brilliant jury-riggers like Paul Padial, a Manhattan social worker who fashioned a rudimentary coffeemaker in the wilderness. And publicist Cynthia Morse, also a New Yorker, who discovered an alternate source of salt when her van got stuck in an ice storm. And Joshuah Bearman, a writer living in Los Angeles who, along with a clever friend, employed gravity to heat up a hot tub and save New Year's Eve. I'm also talking about evil geniuses like Vince, the antihero of Francine Maroukian's story, an actor whose cheater's mind and sleight-of-hand saved the two-timing author from a mighty awkward situation. And Vincent O'Keefe, a stay-at-home dad from Toledo, Ohio, who takes a truly emasculating step to coax his baby daughter into taking a bottle. And Chris Jones, a sportswriter from Ottawa, Canada, whose simple but brilliant strategy helped him track down running back Ricky Williams when Ricky grew bored with the NFL and ran away—all the way to Australia.

These last few contributors didn't *invent* anything—their stories involve no duct tape or Swiss Army knives or half-melted Milky Way bars—but in my book, they absolutely qualify as MacGyver stories. Wikipedia, the open-source online encyclopedia, states in its "MacGyver" entry that to pull a MacGyverism is to "fix, repair, rig, solve, build, invent, or otherwise save the day." It's that last part—saving the day—that all these stories have in common. But *What Would MacGyver*

Do? takes Wikipedia's definition and broadens it even further. This book celebrates acts of improvised genius, period.

Shortly after Richard Dean Anderson got the MacGyver job, he told the *Akron Beacon Journal*, "I'd been turning down a lot of things for the last year or so. I'm trying to let integrity be an integral part of my personality. This character has a lot of the qualities that I've been looking for. He's a very physical character, (but) there's a humanity about the character that is very attractive to me. He's not relying on an underlying vein of machismo to get through all this. I'm going to embellish the hell out of this character. They have no idea how well they cast this."

Well, Richard, you really nailed that one. For many *MacGyver* fans, the character's fundamental goodness goes to the heart of his appeal: In a dark and stormy world, Angus MacGyver can always be counted on to do the right thing. But to the younger generation of MacGyver fans—to those twenty- and thirtysomethings who vaguely recall the TV show from their childhood but were reintroduced to Mac though hip-hop and *The Simpsons*—Mac's appeal is more about his resourcefulness, his unflappable cool no matter how high the stakes. His kitsch value is also part of the equation, and, yes, so is his hair. In editing this collection, I've done my best to include stories that will appeal to the full range of MacGyver fans—fans of the show, fans of the man, and even fans of the *idea* of MacGyver.

One last thing. In the About the Book section of WhatWouldMacGyverDo.com, I wrote the line: "We've all pulled a MacGyverism or two in our day." Actually, that's not true. Many of us have *never* pulled a MacGyverism—not even once. I had every intention of coming up with a story for this collection, despite my incompetence in the improvisational arts. So did my editor at Hudson Street Press, who came up with the idea for the book in the first place. But after wracking our brains for six months, neither one of us could come up with a single story that seemed worthy. (We'd both found ourselves in plenty of tight spots, but the MacGyver moments were lacking.) And that's because it's *hard* to be this clever. It's hard to be MacGyver, if only for that one shining moment. Especially in real life, when the only script is the one you write for yourself.

So to all you contributors, including those who submitted stories we weren't able to publish, I thank you. My editor thanks you. Hudson Street Press thanks you. And most of all, MacGyver thanks you. Wherever you fall on the continuum of MacGyver fanhood—whether you own all seven seasons on DVD or simply love the man for his mullet—we hope you have as much fun reading this book as we had making it.

Oh, and if you ever lose your bookmark, you know what works pretty well? A paper clip.

THE JUNK IN
THE TRUNK

EVERY DAY, ALL ACROSS AMERICA AND AROUND THE WORLD, drivers hold their cars together with little more than love, duct tape, and desperate invention. This chapter is for them (and for those of us who can learn from their resourcefulness).

In most of this book, the definition of "MacGyverism" is loose by design. Almost any act of improvisational brilliance can qualify, so long as it demonstrates quick thinking, gets the author out of a tight spot, and caps a good story. But the pieces in this chapter are classic MacGyver stories, characterized by ingenious solutions using available materials. Angus MacGyver would be impressed. In fact, I'll go so far as to say that he might even learn a thing or two.

Take, for example, Amber Adrian of San Francisco, and her story about locking her keys in her car. I now keep a ball of twine in my glove box. Or Issa Eismont's tale of repairing a blown clutch pedal in the Nevada desert. I probably couldn't replicate his brilliant fix even if I had access to all the same materials, but my wife does knit, so at least I know I'd have a chance. Or former frat guy Harry McCoy's story about protecting a tire from a malevolent fender. More so than in any other chapter, these stories contain lessons that are genuinely useful.

Why is the car such a hotbed of ingenuity? Two main reasons, I think, the first of which is obvious: because so many things can go wrong, from minor inconveniences (a radio knob that won't stay on, a glove box that won't latch shut, a broken windshield wiper) to life-threatening malfunctions (a blown tire at eighty mph, an errant air bag, or a hood that pops open at exactly the wrong moment).

The second reason is less obvious but arguably more important: because people let so much crap pile up in their backseats. Old Chalupa containers. Cracked CD cases. Chewed gum. Unchewed gum. Of course, this kind of random debris is what great MacGyverisms are made of. In fact, I think it's safe to say that people who keep their cars perfectly clean are significantly less likely to be able to pull a roadside MacGyver—and much *more* likely to be forced to stand around on the shoulder, waiting for AAA.

Speaking of AAA (and other organizations that offer roadside assistance), I'd like to extend a special thank-you. If you were always quick to the scene—or even if you lacked your reputation for arriving a tad *late* to the scene—I probably would have received far fewer submissions for this chapter.

Amarillo Ice

by Cynthia Morse

THE LAST TIME I'D BEEN TO TEXAS, IT WAS HOT AND STICKY AND everyone wore either big hats or big hair to keep the sun at bay. So I was a little disoriented when I pulled my white, mud-speckled Dodge rental van off I-40 into the icy city that claimed to be Amarillo. This wasn't Amarillo; it was Cryogenically Frozen Amarillo, and I was clearly a foreigner to the ways of the Ice Cowboys.

It was dusk, and the rush-hour traffic was heavy. Weren't these Amarilloans confused by the ice on the road and the snow on the rooftops? Apparently not. They all seemed to be driving with confidence, making me even more nervous about plowing into them. Yes, it was January, so perhaps I should have been prepared. But this was Texas! The state's reputation for warmth and sunshine was why I'd taken this indirect route for my postgraduation move from Northern California to New York City in the first place.

At this moment, however, I wasn't exactly feeling the warm glow of safety as my van jerked and swerved and tried to change lanes without my permission. After an unscheduled side trip to an empty parking lot (to turn around) and a residential street (to threaten parked cars), I decided to get off the road ASAP. I somehow managed to steer the van into the parking lot of a respectable-looking motel with a red neon VACANCY sign. I glided into a space near the office, and a few minutes later I had a key for a room all the way across the vast parking lot and around the corner. I started the engine, threw the gear shift into reverse, and attempted to back the van up.

The van did not back up. The van did not move at all. The wheels spun merrily on the inches-thick ice and showed no intention of ever changing their plans.

Perhaps this was why so many people had been so averse to the idea of me traveling alone? If this had happened in Seattle, where I had grown up, I would have dialed the numbers of a few strapping young men and asked them to push me to freedom. But the only person I knew in Amarillo was the tiny, elderly, kind-eyed woman who'd just checked me in to the motel—and her hands had shaken with effort as she swiped my credit card through the machine.

What now? I couldn't leave the van there all night, far from my view and filled with everything I owned. And besides, I didn't want to haul my bags that far in the cold. Panic began to creep in; I fought it off. I will figure this out, I resolved. But first I needed a snack.

Yes, I am what you might call a "stress eater," and this moment certainly qualified as stressful. Still sitting behind the wheel, and with a hint of desperation, I reached across the passenger seat and grabbed the colorful gift bag that a friend had filled with snacks for my long drive. This stash had helped me survive the lonely hours on the road, one gummy bear at a time.

But it wasn't gummy bears that my hand found this time. It was a gigantic bag of Chex Mix. I tore it open, grabbed a handful, and shoved the salty bits in my mouth as I stared out the windshield at my ice-covered surroundings.

I thought again about what I would do if I were back in Seattle. Well, I wouldn't be in this situation back there, I reasoned, because I always kept some cat litter in my trunk during the winter months to put under the tires if I was ever in a jam. If only I had something just as gritty, something that would work just as well as—

I froze, mid-bite.

I admit, I hesitated. I do love Chex Mix. And the odds were pretty good that it wouldn't work, in which case I would be not only stuck in the cold with a van that wouldn't move—I would also have wasted some valuable snacks.

But I was feeling emboldened now that I had an actual idea, and one that didn't involve knocking on the doors of the other motel guests. I hopped outside the van and sprinkled the mix on thick, as close to the tires as I could, moving tentatively in my tractionless canvas shoes from one tire to the next. I was trying to finish before the cold cut through my flimsy sweatshirt, but I also wanted to do a thorough job. When I was finished, I stepped back for a moment to admire my handiwork, noting how ridiculous I would have looked if there had been a single person within mocking distance.

I leaped back into the driver's seat and started the engine, then turned up the heat and revved the engine a few times, giving the Chex Mix some time to settle into the ice. I put the gearshift into reverse once more and stepped on the gas, hoping for a miracle.

At first, the tires just spun in that now-familiar way, and I felt a little sinking feeling. I stepped off the gas. I had no idea what to do next. As tears of helplessness started to well up in my eyes, I jammed the accelerator out of sheer frustration. This time, though, I felt a rocking sensation as the tires seemed to get some purchase. To my growing excitement, the rocking gained momentum as I inched back and forth, and suddenly I was rolling backward out of the parking space. I drove across the parking lot and around the corner to my room with the giddy, empowered feeling of the self-sufficient human.

As I drifted off to sleep in my warm bed, I made a mental note to ask at the front desk in the morning if there was a supermarket within walking distance. I was out of Chex Mix, after all, and that spot where I just parked in front of my room looked suspiciously slick.

Cynthia Morse moved from California to New York City several years ago because she graduated and couldn't think of anyplace more expensive to live. She works for a restaurant public relations agency and shares an apartment with her cat, Captain Morgan.

Madame Defarge Saves the Day
by Issa Eismont

IT WAS ONE OF THOSE TRIPS THAT SEEMED JINXED FROM THE START. We'd postponed it several times for several reasons—the dog-sitter was unavailable, my wife's boss reneged on previously approved vacation time—but now Amie and I were finally on our way to a well-deserved (if budget-minded) vacation. It was a sunny and seasonably warm October weekend in the Bay Area, and we looked forward to a straightforward nine-hour drive to her parents' house in Boise, Idaho. Not the most romantic destination, granted, but we were desperate to escape the daily grind, and the price was right. We loaded up our 2002 Ford Focus and hit the road at 7 A.M. on Friday morning.

Around noon we stopped for lunch in Winnemucca, Nevada, a sleazy little town of two-bit casinos and cheap motels. Having made this trip many times before, we knew two things about the remaining two hundred desert miles between Winnemucca and Boise: 1) there was no cell service, and 2) there was nowhere to eat. So we enjoyed a gourmet meal at a truck stop (chili dog with cheese for me, beef jerky and a Diet Coke for the missus), stocked up on bottled water, and called the in-laws to let them know we'd be there in four or five hours.

It was Amie's turn to drive. Just outside of Winnemucca, as we crested the hill beyond which cell service disappears, she made a face and muttered, "The clutch feels weird." Then, almost immediately: "The clutch is gone!" As she pumped her left foot up and down, trying to get the clutch to react, I heard the hollow, harrowing sound of the pedal hitting the car's carpeted floor mat. There was zero resistance.

She shifted into neutral and turned the wheel slightly to the right. As the Focus coasted to a stop on the shoulder of the desolate highway, the only sound was the wind blowing sand and tumbleweeds across the bleak landscape.

Oh, and from Amie: "*Shit.*"

Fortunately, we were only a quarter mile into the no-phone zone. We walked back till we had reception and called Ford roadside assistance. Their response was less than ideal. Because of our remote location, it would be at least two hours before they could get a tow truck to us.

At this point my wife, who never cries, was shaking with frustration; she looked like she might explode.

Okay, time to do *something*. We ran back to the car and I contorted myself into an impossible position on the floor of the driver's side to see if I could figure out what the hell had gone wrong. I spotted the problem immediately. The metal pin attaching the clutch pedal to the actuator (a small plastic thingamajig under the pedal that continues under the floorboard, and that is a key part of the larger mechanism that engages the transmission) had completely severed. The pin was rolling around on the floor of the car and couldn't be reattached without a welder. Now, I've worked on my fair share of motorcycles and know my way around cars, but I'm not in the habit of traveling with a welder in my trunk. On the bright side, the weld had failed cleanly, leaving a perfectly round hole in the clutch pedal. All I needed was something to fill the hole and serve as a replacement pin. Then I'd be able to reconnect the pedal to the actuator, giving the pedal the ability to engage and disengage the clutch and actually allow us to, you know, maybe shift gears.

I walked around to the back of the car and began rummaging through the trunk, waiting for my *Eureka!* moment. At first, nothing. But then my eye fell on one of the incredible number of bags Amie insists on traveling with: her knitting bag. Hmmmm. I rooted through it until I found a needle that looked about right.

"Honey," I said, "how attached are you to this knitting needle?"

Now, while I'd been trying to figure out the problem, Amie had walked back up the road and called Ford again. They repeated that it would take at least two hours for a truck to get to us, but this time they informed her that once the truck arrived, it would need to tow us *back* to Reno—two hours in the wrong direction. So Amie was now a ticking time bomb of rage.

"Not as much as Madame Defarge was attached to hers," she hissed.

Somehow the reference to the bloodthirsty knitter from Dickens's *A Tale of Two Cities* didn't comfort me. But if I could just pull this off, all would be right with the world.

I folded myself once again onto the floor of the driver's side. My back and neck voiced their objections. I was definitely going to need a chiropractor the next day, but first things first. I held the needle next to the pin and compared the diameters. They were nearly identical. I threaded the needle through the hole in the clutch pedal and the hole in the actuator, then maneuvered the actuator up the needle. I snapped the excess needle off, leaving just enough room to bridge the gap between the pedal and the actuator. The head of the needle would keep it in place, preventing slippage.

"I think this is going to work," I muttered, though not loud enough for Amie, who was pacing frantically outside the car,

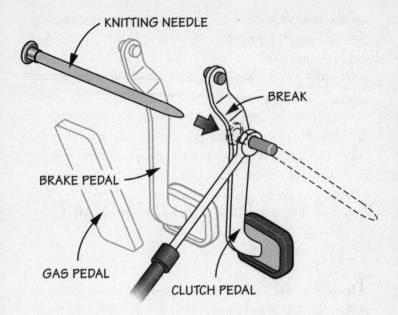

KNITTING NEEDLE

BREAK

BRAKE PEDAL

GAS PEDAL

CLUTCH PEDAL

to hear. I uncontorted myself from the floor of the car and sat in the driver's seat. Amie climbed in the passenger side. I started the car and gently depressed the clutch pedal. Tension! (Though maybe not as much as the tension in the air.) I stepped on the gas and slowly moved along the shoulder until I had enough speed to attempt to shift. I don't think either one of us breathed until the clutch pedal completed its journey the floor, allowing me to shift smoothly into second.

"IT WORKED!" I shouted.

Amie's face lit up into a huge smile. "You are *soooo* my hero!"

One great big kiss and a few minutes later, we were sailing along at seventy-five mph. We called Ford and told them we no longer needed their services. (They seemed unimpressed.) And we made it to Boise by cocktail hour.

Issa Eismont lives in Oakland, California, with his wife and two dogs. They've made several incident-free trips to Boise since the one described here, probably because they now take their motorcycles instead of the car.

"¿Cómo se Dice 'Acetylene Torch'?"
by Stewart Engesser

THE TROUBLE STARTED AFTER MY FOURTH MOJITO. OR RATHER, IT started the next morning, as I merged shakily onto the Henry Hudson Parkway in New York City, already late for the modern poetry class I was scheduled to teach at one of New Jersey's fine state universities. As I attempted to navigate my way through rush-hour traffic, the exhaust system of my old brown Mazda sedan seemed to suddenly explode. There was a crashing bang, and then something large, something metal, fell to the asphalt beneath the car and began to drag. I glanced up at the rearview mirror and noticed a merry shower of sparks dancing on the road behind me, nicely complementing the bright fall foliage in Riverside Park. As the car roared and coughed, I coasted to the narrow shoulder, traffic swerving around me, horns blaring, middle fingers waving.

This was going to be a problem. I am not mechanically inclined. Nor am I "handy," or "improvisational," or "clever." I can barely navigate my way through a toll booth. I'm a hopeless, disorganized dweeb with a large record collection, bad posture, and a taste for books about disastrous polar exploration and the horrors of trench warfare in World War I.

But I do have an imagination.

So I sat in my broken car and imagined where this turn in my morning might lead. Poverty? Unemployment? Being abducted and eaten by death cultists? Anything seemed possible. I was still dressed in the blue and white seersucker suit I'd worn to the party the night before. The pants were stained with mojito and cigarette ash, and the jacket was rumpled from lying in a jumbled heap on someone else's bed, with me in it.

I tried to focus my thoughts. What would my father do? He was an engineer. He knew about cars, and could even identify specific parts, such as the "valves," the "oil pan," even the pleasantly eccentric sounding "crank case." If only I had paid attention at all those Saturday "let's learn about the wonders of the four-stroke engine" bonding sessions when I was a teenager. But I hadn't. The truth is that my father's logical, facts-based world bored and antagonized me. If he had been telling me how Link Wray got that dirty fuzz tone guitar effect in his seminal 1958 instrumental classic, "Rumble," I would have listened. But this was an area of interest that we didn't share, so now, stranded on a dirty stretch of road, I had no idea what my father would do, although whatever it was would have probably involved using some of the many tools he always kept in his trunk.

My "tool kit" was comprised of a single, bent, wire hanger

that my friend Harry* had left in my car after a party in Brooklyn. I didn't even have a functioning cell phone on me— I'd spilled coffee on mine a week earlier and hadn't gotten around to ordering a new one.

I needed a more appropriate role model. Someone used to making do with less. Someone like . . . MacGyver. Well, thanks to Harry, I had a wire hanger, and that was a start.

I'd driven enough junked-out cars in my thirty years to know that when the exhaust system drops, you either a) tie it to the frame of the car with whatever's available, such as a tube sock or a shoelace, and head for the nearest mechanic, b) walk away after destroying the vehicle so that no one can trace it to you (something I've found especially easy to do in places like Australia, where no one particularly cares when they see someone pushing a car off a cliff), or c) call AAA.

My AAA membership, purchased in flusher times, had long ago lapsed. (Not that I had a phone to call them with anyway.) And, unfortunately for me, New York City is nothing like the rugged coastline south of Perth.

So it looked like Plan A.

The Henry Hudson Parkway runs along the Hudson River up the west side of Manhattan, past Riverside Park, toward the George Washington Bridge and the Bronx. It's an old, narrow highway, and cars often travel at speeds in excess of five thousand miles an hour. As I got out and began to take stock of my situation, I discovered something disturbing. A lot of trash gets blown into the gutters of the Henry Hudson—and now I was part of it. Food wrappers, socks, one gold stiletto, sand,

*Names of partners in crime have been changed to protect their privacy.

grit, cigarette butts, and . . . great God, is that a human scalp? No, just an old wool hat wrapped in seaweed.

I peeked under the car and, sure enough, the exhaust pipe had broken off from the rear of the muffler, and now was bent and deeply wedged above the axle, with one ragged end resting on the asphalt. The vehicle would still run, albeit with the throaty, chugging roar of a Formula One race car. But I would need to ride with all the windows down to avoid death from carbon monoxide poisoning. And with the pipe dragging like that, I was afraid the sparks would set something on fire. Something like the gas tank. The only thing to do was tie it up—or cut it loose.

After pawing around under the maroon faux-velvet seats, I came up with a rusty pair of pliers, jumper cables, sixty-seven cents, several books of matches, a Perry Como mix tape that had belonged to the car's previous owner, and my trusty wire hanger. I could wrap the hanger around the busted pipe, using the pliers, then somehow fasten it to the frame. Or maybe I could wrap the jumper cables around the pipe and tie it up that way. Better than using my sock. I looked at my watch. I was now quite late for the class I was supposed to teach, and all indications were that I was going to miss my next class, too. My students would be thrilled. I might be fired. Whatever fix I was going to come up with, it needed to be quick.

That's when I saw the Guatemalans. They were headed straight for me. And I could tell, mostly by their guttural shouting and angry, demonstrative hand gestures, that they were pissed.

There were five of them. They wore hard hats and work clothes—clothes that fit this hazardous, traffic-blasted landscape in a way that made my vintage seersucker seem even more fey and ridiculous than it had the night before. Before going out last night, I should have stuffed a change of clothes into my messenger bag. Of course, you never think of these things when you're heading out for the evening. You never think, *Hey, I bet I'll end up having a few rum drinks tonight and making a pass at an old friend, and miracle of miracles, get semi-lucky, and end up sort of fooling around with her on her parquet kitchen floor while the German-language version of David Bowie's "Heroes" plays on repeat on the CD player.*

Well, sometimes you think of those things. But I'd been in a hurry. And now the no-nonsense Guatemalans were getting closer. Why were these people yelling at me? I was the one who should be yelling. My car is broken! I'm late for my class on T.S. Eliot's *The Wasteland*!

Luckily, I studied Spanish in New Jersey public schools for eight years. So I was quickly able to deduce what the Guatemalans were so hot about. They wanted me to move my car because I was parked in a folk-song zone. *Construción.* That's the word for "song," right?

"I can't move my car," I tried to explain. "It's broken, like my Spanish."

But they kept yelling. Then I had a brilliant thought. These were working guys. I bet they had tools. I made a cutting motion with my hands. *Do you have wire cutters, like for cutting bike locks, or chain-link fences?* They seemed confused. I looked down at the hand motions I was making: bringing my closed hands together and apart, together and apart, simulating the motions of using a large pair of wire cutters. I saw that it made me look like one of those toy monkeys, the kind that bangs a pair of tiny cymbals together as it sputters around the floor. Giving up, I pointed weakly under the car. *"Por favor,"* I said. "Look beneath the auto."

One of them took a look. "I cut!" he yelled. "I cut!"

He hustled over to a pickup truck parked fifty feet away and came back with what appeared to be some sort of flame-thrower: a metal wand, bent at the tip, attached by a hose to a large cylindrical tank. Once he got close, I recognized it as something I'd seen in art class back in college, during the sculpture section. It was an acetylene torch. Who knew it had an actual practical use? He rolled under the car, and in ten seconds he'd lit the torch with a mighty *whoof,* cut through the pipe, and yanked it loose. He got up and dusted himself off, then tossed the pipe into the trash-filled gutter. "Now go!" he yelled.

I got in. I started the car, which immediately roared to life. *So this is what cars sound like with their muzzles off,* I thought: *giant roaring hell dogs.* I could already smell the exhaust fumes. If I kept all the windows down and reduced the frequency of my breathing, I might just make it as far as Teaneck.

I rolled down the driver's-side window and held out some money to pay the Guatemalan for his trouble, but he wouldn't accept it. He pitied me, I think. An overgrown child, far from home, dressed for an ice-cream social. At least I'd had the presence of mind to remember that age-old lesson: Never jerry-rig something if you can cut it loose and abandon it forever along the Henry Hudson Parkway.

Stewart Engesser is a television and advertising writer, and the creator of Mysteries of Science Explained, *a vidcast available at mysteriesofscience.com.*

All He Needed Was an Umbrella Spoke and a Length of Green Twine

by Amber Adrian

WHAT'S THE FIRST THING YOU WOULD DO IF YOU HAD JUST LANDED in a dusty parking lot on a hot September day, eager to run rapids and spiral down waterfalls with your boyfriend and fourteen other hardy souls? You, my non-accident-prone friend, would likely grab an inner tube, pop open a beer, and wait for the fun to begin. I, on the other hand, slathered on some sunscreen (missing several key places, I would later discover), locked my car, and slammed the door shut—with my keys sitting on the drivers' seat.

That's right. In the outback of Sonoma County, California, two hours from my San Francisco home, I locked my keys in my 1999 Suzuki Esteem.

John and I had just arrived on the banks of the Russian River for Flotilla 2005, our annual festival of pure hedonism. It was time to submerge scantily clad butts in the chilly water and float lazily downstream under the Northern California sun, stopping occasionally for cold beverages and grilled meat. This agenda did not, however, allow for disastrous misadventures of the idiotic persuasion.

While everyone else was bustling about, angling for the best inner tubes and wondering how to waterproof their cell phones, John and I called the AAA hotline. Halfway across the world, a phone rang in Bangalore. The man who answered was very polite. He tried hard to be helpful, but he had no idea where we were, probably because neither did we. Apparently, telling AAA to "look for the bendy place in the highway near the exit with the clump of trees and all the cars" isn't enough for them to go on. Bottom line: AAA wasn't coming.

The situation called for resourcefulness and ingenuity. Since I'm neither resourceful nor ingenious, I started looking for a rock to heave through the window. Luckily, there were brighter folks around.

John began gamely, even cheerfully, searching the scrub brush at the edges of the parking lot for a coat hanger to jimmy open the door. I tried to help. We found a stray boot, a hypodermic needle, and lots of rocks, but no handy wire hanger. So John dug through his backpack and pulled out his Leatherman tool (basically a Swiss Army knife on steroids), then started to cut a piece out of a nearby barbed-wire fence.

Meanwhile, another plan was afoot. Paul, a good friend from previous Flotillas, was inspecting the car's round manual door locks, which, when locked, stick up about half an inch from the door. He muttered something about string. I didn't catch what he said, but he sure seemed excited. Soon his muttering turned to rummaging. He opened his car (which was conveniently unlocked) and scrounged an umbrella and a length of green twine from the back seat. Using John's Leatherman, he cut a spoke from the underside of the umbrella. Then he threaded one end of the twine through the hole at the tip of the spoke and tied it off. At the other end of the twine, he tied an open knot, creating a loop about an inch across, and leaving a long tail of loose twine.

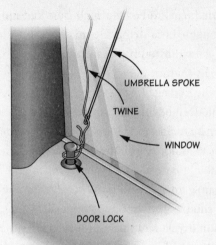

UMBRELLA SPOKE

TWINE

WINDOW

DOOR LOCK

Amidst much naysaying from Team Flotilla, Paul, now the official Patron Saint of Idiot Girls Who Lock Their Keys in Their Cars, turned toward my Suzuki. He had a determined look on his face. In fact, he looked ready to kick some locked-car ass.

As he and John discussed the likelihood of fitting the rod through a space designed to keep out rain, John unfolded the pliers from his Leatherman and gingerly pried the top of the window about an eighth of an inch away from the door frame. Paul slid the umbrella fishing pole through the crack and maneuvered the "line" just above the lock, keeping the loose end of the twine hanging out the top of the window. He then tried to loop the open knot over the lock button. And completely missed. He jiggled the string. He missed again. He jiggled it a third time. And a third time he missed.

"$&#@!" he said.

"$&#@!" he said again.

He took a deep breath and gave it one more shot. And this time, finally, he managed to loop the knot over the lock. Then, very slowly, so as not to pull the loop back over the button, he tugged on the loose end of the twine, closing the knot snugly around it. And finally, God bless him, Saint Paul pulled the string taut and popped that puppy open.

As he would yell multiple times throughout the day and well into the night, "I MacGyvered the hell out of it!"

Amber Adrian is a theater writer in San Francisco who blogs about her supreme ineptitude at mooseinthekitchen. blogspot.com. She is sorry to inform you that all the stories are true.

The Little Wrenches That I Couldn't Believe Actually Could

by Harry McCoy

IT HAPPENED ON A DISMALLY COLD, RAINY SUNDAY IN JANUARY OF 1966. Not only was the weather miserable on the Jersey Shore, but my head wasn't feeling so well, either. I was home on semester break during my junior year of college, and I'd spent the weekend bar-hopping in New York City with old high-school friends. I had just settled down on a well-worn sofa in my parents' family room for an afternoon of football when the phone rang. I considered not answering it—I truly had no desire to do any more socializing that weekend. But for some reason I picked it up. It was Bill, one of my fraternity brothers, and he needed my help.

You know those guys who just seem to have a black cloud over their heads? Bill is one of those guys. On this particular occasion, he had spent the weekend as I had: partying in Greenwich Village. He'd had one too many beers the night before and had jumped a curb on East Sixth Street, blowing out his tire and ending up with a bent fender. He'd spent all his money in the bars, too, so a tow was out of the question. And since this was 1966, an ATM was not an option. To make matters worse, two burly NYPD officers were not-so-patiently waiting for him to move his car, a red '64 Plymouth Fury, which was jutting out into the street. Not to mention that, at any minute, the rain was predicted to turn into what the *New York Times* likes to call a "wintry mix," i.e., freezing rain, sleet, and snow.

"Could you please come up and help me move this piece of crap?" he pleaded.

As any handyperson knows, handiness is both a blessing and a curse. A blessing for obvious reasons: You save a lot of money fixing stuff yourself. But a curse because once everyone *knows* you're handy, you get calls like this all the time. I was the resident Mr. Fix-it of Phi Delt, a fact not lost on Bill.

There was nothing I wanted to do less. But what could I say? He was a brother—and the abuse I'd get back at school if I didn't step up would be merciless. So I grabbed my father's toolbox from the garage and grumpily headed out the door and up the Garden State Parkway to New York. An hour later, I found Bill, shivering by his car in what was now a steady rain. As the cops looked on, increasingly impatient but apparently not enough so to lend a hand, we changed his tire, only to discover that the fender had bent in such a way that it was now rubbing against the spare. If Bill turned his steering wheel to the left, even just slightly, the spare would be abraded and blow out. And we still had no money for a tow. And the cops were still waiting. And our heads were still throbbing.

Groping for an idea, I started rummaging through the toolbox. I knew I needed a "spacer" to separate the tire from the bent fender, which we had no way of straightening. A piece of wood? A piece of metal? The tricky part was the thickness. Too thin and it wouldn't create enough space between the inside of the wheel and the brake drum, which could cause a blowout. Too thick and I wouldn't be able screw the lug nuts on tight enough, which could cause the spare to loosen (and maybe even come off) in transit. Either way, not good.

Two open-ended wrenches were the best I could come up with. (Did I mention that the rain was starting to freeze?) We removed the spare again. Using black electrician's tape, I taped the wrenches onto the brake drum between the lug bolts. Then I put the spare back on and started tightening the lug nuts. As I tightened, the extra cushion of space created by the wrenches pushed the tire outward just enough so that when Bill turned the wheel, the bent metal missed the tire. Alleluia! In theory, Bill should now be able to drive—and steer—without blowing the spare.

We got in the car, eased off of the sidewalk, and gingerly hobbled several blocks. Then I jumped out to have a look. It seemed to be holding. Bill then cajoled me into following him through the Holland Tunnel into New Jersey. It was dark now, and the freezing rain was mixing with sleet. We stopped again on the Jersey side. And to our utter surprise, my solution was still holding up.

Bill decided we could make it back to his parents' house in Manasquan, not far from my folks' place in Point Pleasant. But we couldn't risk a speed of sixty-five miles an hour on the highway, so we made our way down Route 35 instead, slogging through every little burg between Jersey City and Manasquan. Miraculously, the wrenches held. And two hours later (twice as long as the trip usually takes), we limped into Bill's drive-way.

Needless to say, this episode only enhanced my rep as the MacGyver of our fraternity (a full nineteen years before the first episode of *MacGyver* aired), which, for the aforementioned reasons, had its pros and cons. But those are different stories for other days.

Harry McCoy, a former schoolteacher, now lives in northeastern North Carolina, where he is known as the Mr. Fix-it of his retirement community.

(Mass)(Velocity)=Momentum

by Rachel Toor

WHEN YOU LOOK AT HIM, AT THE MANICURED BEARD AND WELL-shorn, though balding, head, the Italian leather shoes and smart-cut suit, it's no longer easy to recognize Jonathan for the geek he truly is. Mostly you can see it in his hazel eyes, now that he's learned to clean his glasses so they're no longer covered with a grimy film. There are still lots of "ums" and "ahs" that punctuate his sentences, but he's gotten better at getting to the point. That Jonathan is a successful academic doctor who is invited to speak all over the world will surprise no one who meets him; that he is responsible for millions of dollars of research grants seems perfectly appropriate. This is clearly an accomplished individual. However, he's changed a lot in the decade and a half we've known each other.

We were originally set up by well-intentioned friends. When I first saw Jonathan, I thought that I should be more careful in choosing my friends. His pants were about three sizes too big, his glasses were cobbled together with duct tape, and there were traces of his last meal on his tie. On the way to dinner, Jonathan drove with his arm out the car window. Not in the casual way that men have, one arm resting lightly outside the car while steering confidently with the other hand; no, the door didn't close properly and Jonathan had to hold it shut to keep it from flying open.

But Jonathan's deftness soon became apparent. Not because he fixed his TV by tipping it forward at a precipitous angle. (This made watching it, for me, a chore, but delighted Jonathan because he had been able to prolong its life.) Not because he once spent countless days and hundreds of dollars designing an insulated house for his pet pig that had a separate "furnace" room, thermostat, and window. Not because if you give him a problem he won't stop until he's solved it, often in inspired and creative ways. It's that Jonathan brings things to the brink of disaster and then, through ingenuity and quickness and physical prowess, he pulls off a spectacular save.

We were moving in together, Jonathan and I, and had leased a fancy house with too many bedrooms in a snooty neighborhood in Durham, North Carolina. Jonathan had been responsible for finding the movers. He'd set up the date, and on the day of the move, we sat at our respective houses, perched on packed boxes and waiting for the movers. They were instructed to go first to my place, load up, move on to Jonathan's, load up, and then unload at our new love nest. Everything was all set.

Except that they never showed. It got later and later and when we finally called, they had no record of our move.

Naturally, Jonathan and I had both waited until the very end of our leases, and we had to be out that night. At the last possible moment, dangerously near closing time, we rented a gigantic truck and, with the help of our mutual friend, Mary, and my less-than-athletic brother, Mark, we worked until the early hours moving ourselves.

The next morning, Jonathan and Mark took my dog, Hannah, for a walk to survey the grounds. Our new home was nestled at the top of a long, uphill driveway. The moving truck was parked close to the house, facing down.

They had been walking on the street and, for who knows what reason, had looked back, at just the right moment, to admire the house. I believe that my brother, when he closes his eyes at night, can still see it: The truck was rolling down the driveway, headed for the posh home of one of our new neighbors. Jonathan had forgotten to set the emergency brake.

My brother, a lawyer, mentally ran through all of the possible outcomes.

"Oh shit!" Mark yelled, his body locked in panic.

"Oops," said Jonathan.

With superhuman speed, Jonathan sprinted up the driveway. The massive truck was gaining velocity by the second; Jonathan knew too well the formula for momentum. He jumped onto the running board along the side the door, but there was nothing to hold onto. He grabbed the side-view mirror, which swung in his hand, and tried to balance himself, surfing toward the neighbor's home.

Then Jonathan realized that he'd locked the door to the truck.

With one hand he clawed the roof of the truck; with the other, he fumbled in his pants for the keys.

Reaching deep into his pockets he came out with a handful of change and the key to the new house. So he opened the truck with a paper clip—just kidding. He dug in once again and this time found the right key, then jammed it in and tried to open the door. But his body was in the way. The house across the street was less than fifty feet away. Jonathan maneuvered himself to the right of the door as my brother gazed on at a scene straight out of some kind of twisted indie action flick where

Paul Giamatti plays a role intended for Bruce Willis: a bearded, balding hero attached to a rolling truck en route to a colossal moment of destruction.

Somehow, Jonathan got the door open. He swung into the driver's seat. In one swift motion he pulled hard—on the windshield wiper lever.

"Oops," Jonathan said.

He grabbed again, and this time he got the emergency brake.

The truck squealed angrily to a halt, having crossed the street, the front wheels stopping mere inches from the golf-course-green grass of the neighbors' lawn.

"Oops," said Jonathan.

Rachel Toor is the author of The Pig and I: How I Learned to Love Men (Almost) as Much as I Love My Pets (Plume, 2006). She writes to support her habit of running ridiculously long races in beautiful places and teaches at the Inland Northwest Center for Writers, the graduate writing program of Eastern Washington University in Spokane, Washington.

HOME IMPROVEMENT

THIS CHAPTER HAS A LITTLE SOMETHING FOR EVERYONE: BRAND-NEW parents, resourceful mothers-in-law, engineering students from Canada, rodent hunters, asthmatics, and so on.

Broadly, though, the stories fall into two categories. If your idea of a perfect Saturday involves several hours at the Home Depot, I recommend:

- "Brilliant from the Heat," a story about constructing a makeshift air conditioner from twenty-four dollars worth of parts

- "How the Father Fixed the Motherboard," in which the author, a brand-new dad, applies a low-tech fix to a high-tech problem

- "All Jacked Up," an inspiring tale of a father, his sons, their sweat, and the miracle of leverage

These three stories illustrate one of the most fundamental ingredients of a quality MacGyverism: an utter lack of fear that, in the process of trying to make things better, you'll only make things worse. It's Dave Murphy's father sizing up seven hundred pounds of crippled deck and refusing to be daunted. It's software designer Matt Wood staring into the guts of his iBook and seeing something other than the abyss. It's college student Geoff Milburn deciding that if he can't afford an air conditioner then, well, he'll just have to make one.

It's guys like these who made me want to edit this book in the first place, mainly because, as I mentioned in the introduction, I am the exact opposite. I completely lack the MacGyver gene. When I see a problem, all I see are more problems—problems that can be avoided by simply calling a professional.

The other stories in this chapter are about home improvement in the broader sense. Maybe they wouldn't necessarily impress Tim Allen's über-handyman character from *Home Improvement* (ABC, 1991–1999), but they do lead to a saner, more livable domestic life. A few of these stories are:

- "The Chuck-it Bucket," Rachel Snyder's story about restoring some semblance of order to her daughter's chaotic kitchen

- Katherine Sharpe's "Exhaling with Maria," in which the author didn't just save the day—she may have saved her roommate's life

- "The Night Dad Dressed in Drag," stay-at-home father Vincent O'Keefe's (admittedly) emasculating story about nursing his daughter

And then there's Chris Kaye's "Sic Transit Rodentia," which bridges both categories. Kaye's story describes a situation we can all relate to (especially the New Yorkers among us), and it's one of my favorites in the book. One caveat: If you like mice, do *not* read this story.

Brilliant from the Heat

by *Geoff Milburn*

I︀T WAS THE SUMMER OF 2005, DURING MY THIRD YEAR OF CIVIL engineering at the University of Waterloo (that's in Canada), and I was miserable.

I lived on the first floor of an old house in the middle of what can only be described as a student ghetto. Rows of old houses surrounded the university, long ago surrendered by suburbia to college students. Nobody could afford an air conditioner, and fans just moved the stifling air around. Sometimes I took a shower at 4 A.M. just to feel the cold water on my skin.

One day, my girlfriend, Emily, and I decided to seek air-conditioned sanctuary at the movies with a few of her friends. Unfortunately, they wanted to see *The Sisterhood of the Traveling Pants*. As we approached the ticket counter, it dawned on me that I would never get the next ninety minutes of my life back. I needed a good excuse—and fast. Earlier in the week, I'd begun pondering a way to build a homemade A/C. I figured now was as good a time as any to give it a shot. I turned to Emily and mumbled something about solving all our problems. She looked mildly annoyed, but not enough to make a scene. "I'll be at the Home Depot," I said, realizing as I said it that the store would also be gloriously air-conditioned. "Enjoy the movie. I'll pick you guys up in an hour and a half." And then I fled.

I was armed with a student's budget and a half-formed concept involving cold water, coiled tubing, and a window fan. If this worked—if I figured out how to keep us cool tonight—Emily would forgive me for running out on the movie. If not, the temperature in the apartment we shared would be the least of my problems.

Twenty-four dollars and an hour and a half later, I picked up the girls and we headed home. I assembled my materials: an old garbage can, vinyl tubing, copper tubing, my roommate's floor fan, and a large package of zip ties—those small, flexible, plastic fasteners used to secure almost anything that needs securing. I began by coiling the copper tubing in a spiral along the front of the fan, securing it with zip ties as I went. I used an X-Acto knife to slice the vinyl tubing into two pieces, which were then attached to the ends of the copper tubing on the fan. One end went to the bottom of the garbage can filled with ice water that sat on my apartment floor; the other hung out my first-floor window.

I went outside and started to suck as hard as I could on the end of the vinyl tubing. My theory was that the elevation difference between the water in the garbage can and the end of the tubing would be enough to sustain a siphon effect. (A siphon allows fluid to drain from one container to another, using only the difference in elevation between the two containers to power the flow. Probably the most common—and illegal—application is using a garden hose to siphon gasoline from the tank of someone else's car. In my case, I was just trying to siphon some ice water through a contraption that hopefully wouldn't flood my room.)

I had been sucking on what felt like a glue milkshake for a

minute or two when a trickle, then a spurt, and finally, to my elation, a smooth stream of water started flowing from the end of the tubing. The siphon had managed to pull water up from the garbage can, through the copper tubing surrounding the fan, and then out my window.

My hope was that the flow of cold water would chill the copper tubing. The fan would then blow the hot, stagnant air in my room across the tubing, cooling the air. And the damn thing actually worked! While it didn't reach the goose-bump level of the movie theater or Home Depot, it knocked ten degrees off the temperature in the room and removed a bit of humidity as water condensed on the copper coils.

I replaced the garbage pail with a water supply fed by a garden hose patched in through the window. The hose could supply much more water than a single garbage can, allowing the system to run longer. But most importantly, as soon as Emily

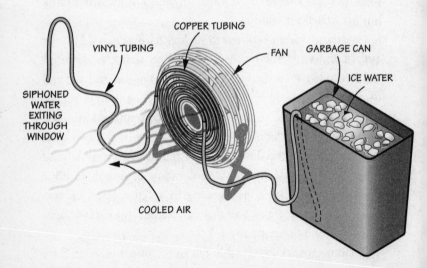

COPPER TUBING

VINYL TUBING   FAN   GARBAGE CAN

ICE WATER

SIPHONED
WATER
EXITING
THROUGH
WINDOW

COOLED AIR

walked into the cooled bedroom, she forgave me for skipping out on the chick flick. Chalk one up for science.

Geoff Milburn wore makeup for the first time in his life when his "Homebrew A/C" attracted the attention of local media, and led to a TV interview about it. For more on his invention, go to www.eng.uwaterloo.ca/~gmilburn/ac/.

The Night Dad Dressed in Drag
by Vincent O'Keefe

EVERYONE'S TRANSITION TO PARENTHOOD IS CHALLENGING, BUT I feel mine was especially traumatic. One day, I was a childless English professor whose most pressing concern was handing out final grades; four days later, my daughter, Lauren, was born a few weeks prematurely. I hadn't even had time to read *What to Expect: The First Year*, and believe me, I knew absolutely nothing about what to expect. Adding to the pressure was the knowledge that in six short weeks, my wife, Michele, would be returning to her OB-GYN residency at the hospital near our home in Toledo, Ohio. She would be the sole breadwinner and I would be the primary caregiver. She worked brutal hours, sometimes more than forty-eight straight, and despite her specialty, we quickly learned that there's a big difference between delivering someone else's baby and taking care of your own. On top of that, we had no relatives in our area to help us.

After three and a half weeks, our problems grew even larger due to the onset of colic, the common but dreaded condition characterized by long bouts of inconsolable wailing. (The baby was often upset during these periods, too.) The only way to comfort Lauren was to breast-feed her frequently for long periods of time—sometimes as often as every two hours—around the clock. Thank goodness Michele was there for those first few weeks.

As week six approached, however, I was getting increasingly anxious. How would I ever be able to console my colicky daughter? Shortly before returning to her job, Michele started to pump her breast milk, and I started trying to bottle-feed Lauren. Unfortunately, she was fervently rejecting the bottle, which of course was my only option as a stay-at-home dad. To calm her, I tried carrying her on my chest for hours, giving her pacifiers, and taking her for rides in our minivan. Nothing worked, at least not consistently. I even tried to soothe her with the sound of the vacuum, which worked pretty well—but only for as long as I kept the vacuum running. After Michele went back to work, my misery intensified. Lauren would cry for hours on end before finally succumbing to exhaustion. But even then, she wouldn't sleep for long. If I didn't find a solution, I was going to need to be committed.

Finally, one night, after an especially horrible ten-hour stretch, I was at my breaking point. My wife had been gone all day and was on call that night, and there I was again: at home and alone with a screaming baby. A new appreciation for the plight of stay-at-home mothers washed over me. As I held Lauren and threw my head back in exasperation, a fuzzy, pink blur in the corner of our bedroom caught my eye. Hanging on a hook near the bathroom was my wife's bathrobe—a very short, plush robe that smelled like vanilla. And at that moment, I got an idea.

In the throes of deep, *deep* desperation, I slipped into my wife's bathrobe. I was aware, on some level, of how absurd this was—but I couldn't have cared less. I sat down with Lauren in my arms and held the bottle against my chest at a breast-like angle. And suddenly, a pause in the screaming. A sucking noise. And finally, blissfully, the sound of a baby beginning her meal. I was stunned. I was elated. Incredibly, this simulation of my wife's scent and shape did the trick. Needless to say, it was not my manliest moment—but it didn't stop me from repeating it in the future whenever necessary.

Vincent O'Keefe is a freelance writer and stay-at-home dad in Lakewood, Ohio. He and his wife have now successfully weaned two daughters.

All Jacked Up
by David S. Murphy*

WHEN I WAS ABOUT TEN YEARS OLD, MY FATHER AND HIS FRIEND built a deck on the back of our split-level ranch house in Warren, New Jersey. It was a good-size deck, perhaps fifteen feet deep

*A pseudonym.

and about eighteen feet wide, designed to maximize party space. The entire perimeter was one large and continuous bench, allowing for ample seating. Throw in a few chairs and you could have a party with thirty or forty of your closest friends. And party we did. No celebration was greater than the dual graduation bash we held in 1986 for my brother Duncan (high school) and me (college).

The problem? The deck wasn't anchored so well, and as we cut the twin cakes (one iced in black and orange, my school colors; the other a vibrant green for my brother's), the damn thing separated from the house and collapsed to the ground—with about forty people on it. It was a six-foot drop, and it happened in three distinct moments: The end closest to the house went first—I remember a spectacular cracking sound as the slim supports suddenly gave way, dropping us down a few feet, which made all the standing partiers (myself included) shriek involuntarily and shift in confused unison toward the house. The back supports went next, dropping us another foot or two and shifting us back out in the other direction, toward the yard. I have an amazingly vivid memory of my Aunt Jeanie lurching forward and crashing into my chest, a look of total disorientation on her face. Finally, the last wooden supports holding us off the ground succumbed to the weight, and the entire deck collapsed to the ground.

It was a short enough fall and, due to the staggered drop, no one was badly hurt. (My future mother-in-law was bonked on the head by a speaker and on the shoulder by a falling beam, which did hurt. She eventually became my ex-mother-in-law, though, so no big worries.)

Obviously, the supporting structure was flawed. The anchor was a two-by-six screwed through a metal plate into the house, which took most of the weight. A pair of two-by-fours had supported the end farthest from the house. This worked fine in the beginning, but over the years, water had seeped through cracks in the plate and rotted away the wood. When all those bodies were on the deck at the party, the weakened wood finally gave up the ghost. But the deck itself was largely intact, a big rectangle of beams that had been nailed into joists; a hobbled giant deprived of its legs, now lying on the ground in my backyard.

My father resolved to salvage it and reattach it to the house.

This time he would support the deck with five concrete footings—one at each corner and another right in the middle. Rather than move the beast, he and my brother pried away a few boards where each support would go and used a post-hole digger (a tool for making deep narrow holes, e.g., the kind you'd need for a basketball hoop) to dig the holes for the tube footings, which they filled with concrete. Then Dad hired a contractor to bolt a sturdy new plate through the side of the house. (Sure, it pained him to pay someone, but this was a serious safety issue, and the last thing he wanted was a repeat of what happened during the graduation party.) The combination of these two things—the strength of footings and the stabilizing effect of the steel plate—would give the deck the support it needed. But they still needed to somehow lift the thing.

I had moved to New York City to embark on my postcollege life, so my father and brother were on their own with a deck weighing a good seven hundred pounds at (more or less) ground level. Duncan had been a linebacker on our high-school football team, and my dad was no slouch in the strength department, either, but it would have taken a young Schwarze-negger and his lifting buddy Franco Columbu to move the thing without help. The plate (through which they would an-chor the deck to the house) had been lowered from its previ-ous position but remained a seemingly unreachable three feet off the ground.

Dad, now a retired restaurateur, is an optimist about many things, especially his odds when faced with a task that would cause most men to seek professional help. But there was no way that he was going to admit defeat to the laws of physics or, worse, pay another dime for assistance on reattaching this deck.

With a steely glint of confidence in his eye, he looked at my brother and gave the order. "Go get all the car jacks you can find."

Duncan raised an eyebrow of concern and confusion, but he set to the task like Sancho Panza to my father's Don Quixote. Soon he had retrieved four car jacks from the garage. Two were mighty 1970s beasts, relics from the days when jacks could lift eighteen-wheelers, or at least a '71 Bonneville; the other two were flimsier jacks from Japanese cars. My dad outlined his plan.

They positioned a jack beneath each corner of the deck and began cranking, using the devices in a way their designers had never imagined. Dad would give one jack a few cranks, then move to the next, cranking that one a bit to try and keep the mighty deck somewhat even. Duncan followed his lead on the opposite side. And like a large, lumbering animal rising from a long nap, the deck slowly rose, inch by inch.

My dad, sweating profusely in the humid Jersey air, wiped his brow and took a mighty slug of Coke every few cranks. Duncan, after initial doubts, began to believe. Dad cranked. My brother cranked. Higher came the deck.

It wasn't until the surface was about a foot off the ground that the peril of the situation began to reveal itself. The deck began to teeter. Seven hundred pounds of wood were now being supported by four car jacks, and things were looking dicey.

But Dad was unstoppable. They kept on cranking, getting that deck closer and closer to its target. When it was even with the footings, a yardstick's length off the ground, they oh-so-gingerly removed the jacks. With the deck now supported by the footings, they were able to rock it gently into its final position. Then they screwed long bolts through the holes that had been drilled through the foundation wall. Inside the basement, my father put an eight-inch-long chunk of two-by-four over each bolt (way too much wood, but that added to the homemade, seat-of-your-pants effect) and anchored them with nuts. A few twists of the ratchet and the deck was secured to the house. A few more screws and the old boards were back in place.

The next weekend, I drove home to Jersey to join my family for Sunday-night dinner. As the sun was setting, we hauled the old gas grill, which had survived the crash, from its temporary backyard home and put it back on the deck, where it belonged. Dad grilled a few steaks, Mom made mashed potatoes, and the four of us dined as a family on our new old deck.

It was a few feet lower than it used to be, and now there was a step from its floor up to the house. But it was solid. Every now and then the old man would stomp his feet on the boards and give us a look and a lift of his brow. His satisfied smile said, *See, I told you I could do it with a car jack.*

David S. Murphy is an insurance executive living in Scottsdale, Arizona. Inspired by his father's unique style of problem-solving, he occasionally walks his dog on a treadmill.

Exhaling with Maria
by Katherine Sharpe

I NEVER UNDERSTOOD ASTHMA BEFORE I MOVED IN WITH MY FRIEND Maria. I thought it was something quaint that happened to old people and the bookish young, and I went on thinking that until one day in the middle of the summer when Maria stopped being able to breathe. The first time it happened we were in a movie theater, watching *Me and You and Everyone We Know*. After a few minutes, the dialogue became punctuated by Maria's struggle to take in air; she sounded like a climber getting sick at high altitude. "I need," she rasped, "to go to the hospital."

Maria had mentioned before that her asthma could sometimes be bad enough to land her in the ER. I'd heard this without really absorbing it. She had one of those inhalers that she toted everywhere, just like other asthmatics. She was twenty-eight, healthy, a painting MFA student at Cornell University.

Everything seemed just fine until that day in the theater, when she hit her inhaler again and again. I heard the *whoosh* of the vaporizing medicine, but it didn't seem to be working. Her breathing dwindled to a tiny, mucousy pant. Not enough air to live on.

I've never driven so impatiently. By the time I led her into the hospital waiting room, Maria was almost crying with panic and frustration. The doctors whisked her away, past the curtains, and hooked her up to a machine pumping vaporized steroids, which steamed like a cauldron of witches' brew. By the time I joined her, she was breathing almost normally, but unable to talk with the hissing inhaler clamped between her teeth. A few hours later, they prescribed her a five-day course of oral steroids and cleared her to go home.

I'd love to say that was the end of her asthma trouble, but it wasn't. A month later it happened again.

It was a hot, humid Sunday at the end of summer in upstate New York. Our friends Yoel and Thomas were over, the four of us chatting and watching TV, when Maria's lungs started to rattle—a scary sound, like a boat hull scraping the pebbly bottom. This time, despondent and breathless, she did start to cry. She began looking for her spacer, which is a plastic chamber that fits onto the end of an inhaler and provides a space for the medicine to diffuse before the patient inhales it. You can take a hit directly, without the spacer (I've seen Maria do this plenty of times), but you don't get a deep enough draw and the medicine doesn't work as well—certainly not well enough to fend off a really serious attack.

Maria checked for the spacer in her purse. It wasn't there. Nor was it in her dresser, her room, or anywhere else.

"Do you need to go to the hospital again?" I asked.

"No!" she bellowed, as much as she could, but it was clear that she just meant "no" to the whole situation—the asthma, having to make a decision about the hospital, all of it. "No! I just need my spacer!"

Worried, trying not to act worried, I did what I could. I loaded her into the car and said we'd go try to buy a spacer. We visited two drugstores and one supermarket without luck, Maria gasping like a goldfish in a plastic bag, the two of us reaching depths of despair that I couldn't have imagined weeks before. "Try your inhaler anyway," I said, and she did halfheartedly, as if knowing it wouldn't help without the spacer, which it didn't.

When we got back to the house, Yoel and Thomas were still there on the front porch.

"Did you—?"

"No."

And that's when desperation gave birth to invention. We needed a spacer, and the stores didn't have one. So? What was a spacer, after all, but a hollow plastic thing with a hole in each end? My imagination scanned the bedroom, the front closet, the kitchen—"Wait! Wait!" I said. "I've got it!" What had popped into my mind was an image of our recycling bin—specifically, an empty bottle of Aquafina that I'd been drinking from at the gym. A hollow plastic thing with a hole in . . . one end. I'd need to make the other one.

After a couple of theatrical flourishes with a power drill, which were ineffective but totally impressed our guests, I hit on a simple solution: Cut an "x" on the side, near the bottom of the empty bottle—the actual bottom proved far too sturdy—with a utility knife.

I looked at Maria. "Give me your inhaler," I said, and pushed it through the "x". "And Yoel, come and hold it here while I seal it up."

Yoel steadied the inhaler so its mouth was just inside the bottle, and I sealed up the gaps between bottle and inhaler with clear packing tape (necessary for proper suction). Then I handed the contraption to Maria, who fit her lips around the mouth of the bottle and released a blast of medicine.

Whoosh.

The inhaler fired; the empty chamber filled with white mist. At my sides, my hands formed eager little fists, as if by clenching tighter I could make it work. And then I saw the look on Maria's face as she breathed the medicine in. Her brows lifted and she nodded slightly, her blue eyes widening with surprise. It was working. We all slackened a little, as though an unseen drill sergeant had put us at ease. Maria flopped down on the couch, exhausted, smiling with the joy of relief. Tomorrow she would go see her doctor and get a prescription for more powerful steroids, but the Aquafina bottle would get her through the night.

I flopped down next to her, also relieved, and high on adrenaline to boot. I squeezed my friend's shoulder and melted into a pleasant haze of pride at my ingenuity, and simple gratitude that for the next little while, at least, everything was going to be all right.

Katherine Sharpe has traded the meteorological extremes of upstate New York for the balmy weather of San Francisco, where she works as a writer and has been known to reuse a single water bottle for weeks on end. She edits a small magazine of first-person stories called Four Hundred Words (fourhundredwords.com).

How the Father Fixed the Motherboard
by Matt Wood

I CAME HOME FROM MY FIRST DAY BACK TO WORK AFTER MY SON Carter was born and discovered that my laptop was courting death. It started leaving me suicide notes, message boxes that appeared every fifteen minutes to say, "Status check indicates imminent hard disk failure." This from my Mac, the better-behaved of the two computers I used daily. This was the one that was never supposed to break, the self-healing, virus-resistant, sleek, sexy sibling of the blocky and brutish Compaq that I used to design software at work. I expected that one to give me trouble; in fact, I had just spent most of my work day restoring files that had mysteriously disappeared during my two-week paternity leave. Now the Mac's hard drive was making a low, growling sound, punctuated by a higher-pitched warbling when I started applications or tilted it just so. I risked going online and ordered a new drive, then started backing up my most important files.

The new drive arrived a week later. But with a new baby and a full-time job, I knew it might be months before I found the time to install it. My wife, Debbie, had just returned to her job as a Realtor, and since we hadn't yet arranged a permanent babysitting schedule, she'd been hauling Carter to her appointments. I decided to take him off her hands for a day and call in sick so I could stay home and resuscitate my computer. This turned out to be a good idea: What I expected to be sixty minutes of work turned into a four-hour ordeal. Apparently, the drawback to having an ultra-compact Apple Power-Book is that its internal organs are packed in tighter than commuters on a rush-hour subway car. I had to loosen two-dozen screws, remove protective foil tape, unfasten wires, and detach the keyboard and mouse conduits before I even caught a glimpse of the ailing drive. One of the final steps was to detach the wire connecting the external power button—the button on the surface of the computer—to the drive. The wire was attached to the drive by a tiny plastic rectangular plug that clipped into a metal connector on the motherboard. I reached through a square-inch gap in the guts of the machine, grasped the plastic plug with my thumb and forefinger, and wiggled it free.

But when I pulled the wire out of the case, the entire connector came with it. I peered into the hole and realized I'd snapped the whole assembly off of the motherboard. There was no obvious means to reattach it; on the bottom of the broken connector were simply two flat metal pieces that looked like they needed to touch two matching pieces on the board. I took the metal portion of the connector off of the end of the wire and tried pressing it back down onto the surface, irrationally hoping it might snap into place, but it fell off and disappeared inside the case. *Arggghhh.* I had to turn the whole computer upside down and shake it like an Etch A Sketch before the piece finally fell out on the table.

I can be very stubborn, especially when it comes to admitting I don't know what I'm doing with a computer. But this time I knew I had messed up, badly. I had no IT department at home, no handy supply of spare parts. And repairing circuit boards was far beyond my abilities.

I was about to start calling around to get quotes on repair costs when I had an idea—a simple, potentially beautiful idea. If I could somehow position the plastic portion of the connector in such a way that those little metal pieces touched, maybe I could fix it. Could I just Krazy Glue the plastic piece to the board? I rummaged through our junk drawer, looking for that magic adhesive, but we didn't have any. So I bundled up Carter and headed to Walgreens. I paced the aisles of the store, carrying my three-week-old baby in his car seat, a look of grim determination on my sleep-deprived face. The other customers looked at me with pity—what they saw was a delirious new father stumbling around the store, looking for formula and diapers. But I found what I needed in the hardware aisle.

I returned home and arranged my instruments on the operating/kitchen table. Using a pair of tweezers, I glued the plastic portion of the broken connector onto the motherboard *just so*, so that the metal portion of the connector touched the metal pieces on the board. Then I reassembled the case, held my breath, and pressed the power button. Amazingly, it worked! With all my sophisticated training as a software designer, with three computer-related college degrees and years of troubleshooting experience, I fixed a two thousand dollar computer with a two-dollar bottle of glue, a pair of tweezers, and one very sleepy baby.

A month after the events in this story occurred, Matt Wood quit his job so he could stay home with Carter and break stuff full-time. He and his family live in Chicago.

The Chuck-it Bucket

by Rachel Snyder*

MY GRANDSON LIAM IS PERFECT IN ALMOST EVERY WAY—SWEET-natured, happy, enjoys attention (but doesn't demand it), and is much sharper than your average fifteen-month-old. (I'm completely objective about this, of course.) He is very well behaved and creates only a minimum of frustration for his parents—with one exception.

Liam's mom (my daughter) is a recipe editor for a cookbook publisher, and it gives her great pleasure to cook for her little darling. He's an eager eater and has developed rather sophisticated taste, with preferences that seem to shift almost daily. Mom goes all out—especially at dinner—to keep the menu interesting and flexible. If, for example, chicken-apple-cranberry sausage with potato croquettes and broccoli doesn't

*A pseudonym.

tickle his palate that evening, there's always spinach ravioli and roasted veggies as an alternative. So in addition to her new role as a mother, my daughter has also become a short-order cook.

But that's not the problem.

No, whatever tension exists between my daughter and grandson stems from *how* Liam indicates his (often fleeting) displeasure, i.e., by hurling his dinner to the tile floor of their Brooklyn apartment. Granted, this is hardly uncommon among toddlers, but Liam seems to take particular delight in flinging his food. Now, it's not such a big deal if he's having, say, penne with a brown-butter sauce. But yogurt with kiwi, on the other hand, is less fun to clean up (especially when it lands in Mom's hair). And the more frustrated she gets, the more fun Liam appears to have.

It's not that Mom hasn't tried to find a solution. First she tried some "age-appropriate reasoning" described in one of her parenting books. Yeah, right! A fifteen-month-old—even a future Nobel Peace Prize winner like my grandson—is just not to be reasoned with, no matter what tone of voice you use. Then she tried to ignore his behavior, hoping he'd stop when he realized no one was paying attention. Strike two. Next she began to hover next to his high chair and attempt to catch the flying food before it actually hit the floor. That worked, kind of, but it wasn't really a practical solution. Finally, out of sheer frustration, Mom decided that removing Liam's food from his high chair until he calmed down was her only option. This, of course, made no one happy, least of all Liam, who often was still hungry.

It pained me to see my daughter so upset at these times, and my adorable grandson so confused and unhappy. What could I do to help? While mulling this over one afternoon, I recalled that, on a recent visit to a pond in the Connecticut country-side, Liam enjoyed playing at the water's edge, placing stones in his little plastic bucket, then carefully removing them—refilling and emptying, again and again. *Hmmm*, I wondered. Would strapping two shallow buckets, one on each side of the high chair, be a solution?

The next time I paid my daughter and Liam a visit, I decided to give it a try. The laundry detergent I use comes in shallow white plastic buckets with short white handles. One bucket was nearly empty, and its replacement was waiting to be opened. So the morning before my visit, I transferred all the powder to a new, larger container and sterilized the now-empty buckets in the dishwasher. Next I rummaged around my catch-all drawer and found two plastic self-stick hooks (left over from another project) to hold the buckets in place. As my daughter and Liam played with Thomas the Tank Engine in the living room, I adhered the hooks to each side of the high chair, hung the buckets from the hooks, and *voila!* Time to see if it worked.

"Lunchtime!" said Mom, and Liam, in typical fashion, ran into the kitchen and asked me to put him into his seat. As he sat waiting for his meal, he was aware that something was different and began examining the buckets. Before anyone had a chance to explain why they were there, he dropped his toy train into one of the buckets to free his hands for lunch. Mom then showed him how he could use the other bucket for the food he didn't want to eat. After a little experimenting—put the sandwich in, take the sandwich out, repeat—Liam seemed more than pleased with himself. He got it!

There'll be plenty of time to explain to Liam why we don't play with our food, or waste it, but until then, mealtime in my grandson's house has become happier for all.

Rachel Snyder is an extremely resourceful grandmother living in Brooklyn. Her grandson calls her "Nanny."

The Extend-a-Rake

by Bruce Hobson

DESPITE MY LONGSTANDING AVERSION TO CLIMBING LADDERS (beyond the first three rungs or so), in 1984 my wife and I found ourselves living in a three-level contemporary home near the Delaware River in New Jersey. This particular dwelling was nestled among numerous mature ash trees, which shaded the house and provided relief from the summer sun. The problem was that these same comforting specimens would then bury the property in an avalanche of leaves each fall. When we bought the house, we had no idea how much work it would take to clear the leaves from gutters, gather the leaves in huge piles, and move them out to larger mountains along the curb for pickup—or that this routine would sometimes need to be repeated every few days.

We moved into our new digs in July, and by mid-October

our decks, lawn, and planting areas were completely covered with leaves. In addition, the gutters were full and the down-spouts were plugged. Clearly, something had to done. Being a logical kind of guy, I formulated a plan of attack:

1. Buy a leaf blower.

2. Get the leaves out of the gutters and onto the two decks. (One was a second-floor balcony along the west side of the house, and the other was a much larger ground-floor deck on the east side.)

3. Move the leaves off the decks and onto the lawn.

4. Gather the leaves on the lawn into large piles.

5. Rake the piled-up leaves onto canvas tarps and old sheets.

6. Drag the sheets/leaves onto the street and make leaf mountains along the curb.

7. Pray that the township would make a leaf pick up run before we ran out of street. (At the height of fall, there was usually only enough room for one car to get through the leaf mountains that lined each side.)

I completed step one without any difficulty—the local hardware store was well stocked with leaf blowers—but then realized that step two implied a lot of climbing up to second- and third-floor rooflines, which, as I mentioned, I don't do.

What I needed, I realized, was a long tool that would permit me to stay on terra firma while cleaning out the gutters. I searched the house, scrounging for parts I could utilize. First I located a sixteen-foot telescoping aluminum pole that I had previously used to touch up the stained siding on the house. Then I found the paint-pad accessory that screws onto the end of the pole and holds the removable pad you use to actually apply paint or stain. This would give me the reach that I needed, but not the ability to sweep the leaves out of the gutters. So I started going through my tool closet and found an old orange clamp that I'd bought years ago and never used. Aha! If I used the clamp to attach a downward-facing dust broom to the paint-pad holder, I might indeed have a ladderless solution to this problem. After scavenging around for a while, I found a short-handled whisk broom in the garage. It seemed about right. I assembled my contraption.

Okay, it *looked* promising . . . but would this thing actually work? I decided to first try the gutter in front of the garage, which could be reached without fully extending the pole. I raised the pole, positioned the brush above the gutter, and gingerly lowered it in. The gutter was seriously clogged, but the brush was sturdy and the clamp held it firmly in place. And soon I was twisting out huge clumps of wet leaves. Oh my!—as Dick Enberg likes to say—it actually worked! After brushing the accumulated debris out of the gutter and off of the garage, I discovered an unexpected bonus. The long, narrow guide bar for the jaws of the clamp projected about seven inches from the paint-pad holder. It could be inserted through the end of the gutter and into the top of the downspout, then wiggled to loosen the plugged section. Once loosened, the clump would soon emerge from the bottom of the downspout, followed by

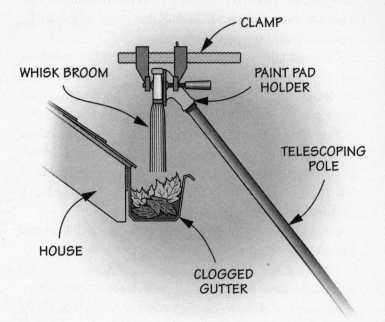

CLAMP

WHISK BROOM

PAINT PAD
HOLDER

TELESCOPING
POLE

HOUSE

CLOGGED
GUTTER

a large volume of nasty-looking water that had been trapped up in the gutter.

Right about then I was feeling pretty darn pleased with myself. That's probably why I was so disappointed when I tried a higher, more difficult gutter that required full extension of the pole—and the top of the Extend-a-Rake fell to pieces. Suitably humbled, I retrieved the parts (which fortunately fell all the way to the ground, sparing me ladder duty) and went back to work. I repositioned the whisk broom for an improved cleaning angle, tightened the clamp, and added a bolt to secure the paint-pad holder. Prototype number two held together, more or less, and I've used it to clean my gutters for seventeen years now.

The contraption never fails to impress, and I've even been asked to build similar tools for others. (Okay, it was my mom and father-in-law who asked.) But the high point of my career as a weekend MacGyver came one day when I was outside putting the tool to use. One of my neighbors drove around the corner, saw me, did a double take, and slammed on his brakes. After bringing his minivan to a complete stop in the middle of the street, he threw open the door, got out, and declared, "That's the greatest thing I've ever seen!"

While I don't think that fellow gets out much, I'll tell you this: It's definitely the greatest thing I've ever invented.

Bruce Hobson is a semiretired safety professional who is often called upon to MacGyver creative (i.e., cheap) solutions to complex workplace hazards.

Sic Transit Rodentia

by Chris Kaye

MY WIFE SAW IT BEFORE I DID. ACTUALLY, THAT'S A LIE. I HAD SEEN the mouse while sneaking a cigarette in my home office late one night. A round, fuzzy little thing. At first I thought it was a Blow Pop that had fallen behind the couch and accumulated a pelt of dust fur. The mouse and I exchanged a meaningful glance, then it disappeared into a crack between the floorboard and the wall.

That it left silently—and that I made no attempt to stop it—sealed our unspoken deal: Don't ask, don't tell.

"A mouse?" I asked my wife, who was now teetering on the arm of a sofa. "Where?"

"Hewentrightthereohmigodhesnuckintothekitchenamouse theresamouseinourkitchenchristopherdosomething!"

Over the next few days I received plenty of orders like this one from Sherri's Mobile Command Center, i.e., any piece of furniture on which she could stand.

"Where?" I'd ask, reaching for my miniature Yankees souvenir bat, which had become my Loathsome Hammer of Rodential Vengeance.

"There—under the desk!" she said one morning. She climbed onto something and pointed. "Oh, wait—it's a moth."

"A moth? You don't know the difference between a mouse and a moth?"

"It was flying around down there where it's dark, so it looked like a mouse."

"Here's a hint on how to distinguish a mouse from a moth," I explained. "Mice don't fly."

"Then how did he get up here?"

Hmm.

Clearly we were dealing with some type of supermouse, a

highly evolved subspecies that could scale heights heretofore unimagined by its fuzzy brethren. For a spineless creature without opposable thumbs, it was certainly crafty.

The question of how this tiny rodent made his way into our fifth-floor Manhattan apartment was a sort of subplot to the main action. Downtime between patrols was filled with speculation: Maybe the mouse snuck onto the elevator, like some Depression-era hobo jumping a cattle car. Wait, do mice swim? He could have come up through the pipes! Or perhaps he was delivered to us in a grocery bag. The man at the corner store was always judging my wife, or so she claimed. Could he have visited some fiendish Korean voodoo upon us that time he delivered beer and ice cream at three in the morning? It all seemed possible.

After consulting with friends, we were offered the services of all manner of pedigreed mouse hunters: dogs, ferrets, a seven-toed mutant cat with paws like thick, furry spatulas, and a boa constrictor named Slash that had singlehandedly rid an entire East Village tenement of rats. My allergies and lack of desire to find a fifty-pound snake dangling from the shower-curtain rod ruled out these options.

We slept on the problem. Lying in bed, every sound morphed into a possible mouse noise. Mice began to sound like the dishwasher. Mice clicked around in high heels from above. Mice howled like the wind.

It was time to call a professional.

I went down to the front desk and asked the doorman for the service-request book, so I could log our problem with the proper authorities.

"What's the problem?" he asked.

"We got a mouse."

His face went white, and he withdrew the book before I could write anything. He clicked on his walkie-talkie and spoke quietly into it.

The super materialized in seconds. I explained the situation, and he responded in a whisper. The building hadn't had any mice in years, he said. *Years.* But he'd take care of it. In the meantime—and he practically passed his hand over my face to complete the Jedi mind trick—*I didn't see any mouse.*

The next day he arrived with an arsenal. Our two-bedroom apartment had an open, loftlike layout with plenty of square footage to devote to low-lying deathtraps. An array of potentially lethal poisons was placed under the sink and inside the radiator covers. Glue traps—thin rectangles of sticky doom—were inserted into tight corners. Traditional old-school neckbreakers were smeared with peanut butter and placed under furniture.

Nothing worked.

A word of advice: Never name something you're trying to kill. In tribute to our mouse's uncanny ability to foil the super, we'd taken to calling him "Mighty." And in naming him, we gave him a soul. He was no longer just vermin, a mere pest. The idea of him skateboarding through the kitchen, hanging ten on a glue trap, was most unappealing, as was the idea of him gnawing off his own legs to escape. And what if we found him still alive in the trap? Then what?

We began to research the most painless ways to do the deed. Drowning sounded like a bad way to go. Sealing him in a Ziploc bag had an S&M vibe that we weren't quite comfortable with. Nothing seemed humane enough, mainly because the endgame of even the gentlest method was, you know, death.

We had to capture him and, through occupational therapy or perhaps the aid of a nice cage and wheel, reform his scurrying tendencies. Or maybe I'd just release him in Central Park, where he could quietly live out the rest of his days until being savagely ripped apart by a roving gang of pigeons.

I set to work building my own PETA-approved contraptions that would satisfy the staunchest of vegans. We had the traditional "propped-up shoe box with string." We had a Habitrail-like device made from empty toilet paper tubes. After studying Mighty's movements, I crafted an invisible wall of packing tape that, if he was running at high speed, he might not see. At the height of my frustration, I briefly considered trying to create a sexy female decoy mouse, with a miniature blonde wig and bright red lips. This was deemed impossible. Both stupid and impossible.

Many of these new traps would be triggered by me, meaning that I needed to be prepared to spring into action at a moment's notice, souvenir minibat at my side in case things got hairy. I spent my evenings seated in the dark of the living room, a lit cigarette in one hand, the string-trigger of my shoe-box trap in the other. This was crazy, sure. Not only that, but I was being outsmarted by an animal with a brain the size of a Skittle. *God forbid this mouse gets ahold of a thimble, some chewing gum, and a can of WD-40,* I thought. *He'll take over the entire building.* Forensic evidence showed that Mighty had indeed entered many of the traps, taken the bait, and crept away un-scathed, after I had dozed off.

After a few weeks, I gave up the game. I put the minibat away and took down my invisible wall of stickiness and my shoe box and my Charmin-brand "Maze of Mouse Mystification®." As long as he didn't let my wife see him, Mighty and me, we were cool.

A few months later, Sherri and I went to Oklahoma to spend Thanksgiving with her family. We relayed the story of her obsession with mouse sounds and everyone had a good laugh at my idiotic attempts to catch him. We watched their dog, a massive black Lab named Henry, chase squirrels around and joked about how great it would be to have him come back with us, if only he could fit into the apartment. My father-in-law, who is from India, related a local legend that involved a mouse being the harbinger of a spirit who wanted to commu-nicate with the living. We were pretty sure our mouse was un-able to speak, but given his other talents, we couldn't rule it out. We ate good food and drank good wine and we forgot all about Mighty.

We returned to the apartment a few days later, and when we opened the door there he was: lying on his back in the entry-way, in a deep rigor, all four legs jutting skyward. Dead. It was almost cartoonish—all that was missing was a black *X* in place of each eye. I scooped him up with a dustpan and dropped

him into a small bag, then took him down the hall to the garbage chute.

Before I dumped Mighty down into the hereafter, I apologized for the poison, which I had completely forgotten about,
and told him how much I'd respected his improvisational skills
(which, we both knew, were far superior to mine). I also asked
him if he had any messages from ghosts in India.

Thankfully, he did not.

*Chris Kaye is a journalist in New York City. He covers technol-
ogy, film, and video games, writing mostly for magazines with
semiclad women on their covers.*

THE TRAVEL CHAPTER

AS I MENTIONED IN THE INTRODUCTION TO THE CHAPTER ABOUT car-related genius, the road is an excellent breeding ground for MacGyverisms. But so is the air. And campgrounds. The supermarkets of France. The mountains of New Mexico. The specific location isn't so much the point. The point is, travel—the experience of simply being somewhere else—often leads to invention and reinvention. Why?

On a practical level, it's simply because when things go sideways far from home, you have no choice but to make do with whatever supplies are available. David Wondrich demonstrates this beautifully when he MacGyvers a round of martinis from an ill-stocked bar at a friend's country home, using "vermouth" made of pinot grigio, rosemary, thyme, and some chamomile flowers.

But the more essential thing you leave behind is the tyranny of your own identity. When you travel, you're not merely some-*where* else. You can be some*one* else, too, if only in your own head, and if only for a little while. No one knows this better than the people responsible for selling the charms of Las Vegas ("What happens here, stays here").

Some of the authors of the stories in this chapter undergo similar transformations. Daniel James, a mild-mannered bank executive from North Carolina, becomes a semisadistic opportunist. Natasha Glasser, a self-proclaimed worry queen, becomes a decisive, swift-thinking guardian of the not-so-friendly skies.

But no one does MacGyver prouder than social worker Paul Padial, whose story opens this chapter. Paul's story differs from the ones I just mentioned in that he doesn't so much undergo a transformation as take full advantage of the natural abilities he inherited from his father, a world-class tinkerer. Suffice it to say that the air was cold, the ground was wet, and Paul *really* needed a cup of coffee.

Not to take away from the inspiring achievements of our other heroes. In magazine editor Tyler Cabot's story, his friend Seth performs with honor mere hours after undergoing a colonoscopy. Josh Bearman of L.A. saves New Millennium's Eve in the California desert. And who knows whether Wil Hylton would have survived the New Mexican heat if he hadn't stumbled on a hidden oasis.

In other words, sometimes you need to get a little lost to locate the MacGyver within.

Good to the Last Sock

by Paul Padial

Back in the mid-'80s, when I was still in high school, I would often escape from my hometown of Tappan, New York, to the woods of Harriman State Park, about fifty miles north of New York City. Lake Sebago, near the town of Tuxedo, was my favorite place to camp, fish, play guitar, party with my friends, and, yes, sometimes endure the hardships of the great outdoors. Thunderstorms. Snowstorms. Sprained ankles. Forgotten matches. Forgotten guitar picks. Forgotten beer. But even when everything was going wrong, we always could count on Dinty Moore stew for dinner and a cup of hot coffee at dawn. Except for one trip when even the coffee was in jeopardy.

There were six of us on that trip: me, my brother, Michael, and four of our friends. Following a night of Heinekens around the campfire, we woke up hungover and cold. (It was May, and the mountain mornings were still chilly.) Eventually, my friend Dave walked to the car to get our glass coffeemaker, the old-fashioned kind you put directly on the burner (or in our case, the coals). I watched him as he carried it back with both hands, like it was the Holy Grail. He had almost reached the campsite, when he tripped over a tree stump and . . . *smash!* Dave was unhurt, but all our rustic cowboy visions of sipping coffee around the morning fire disappeared like smoke.

I grew up in a Puerto Rican household, where coffee was an essential part of the morning routine. And my father was a classic tinkerer, always trying to repair things around the house with unconventional materials and tools. As we stared down at the minefield of broken glass lying in front of us, my brother and I realized that we were trained for this very moment. We started brainstorming, asking ourselves, *What would Dad do?*

It was Michael who suggested the sock. Everyone thought he was kidding; I actually recall a few lame jokes about stinky coffee. But I knew he was dead serious. And the idea was brilliant.

First we needed to fashion a filter basket. Being a graduate of my father's School of Wire Hanger Technology, I ran to my car and grabbed a hanger. (Doesn't everyone keep a hanger in the car?) I twisted the bottom half of the hanger into a four-inch circle, about the diameter of the coffee cups we were using, and bent the other half into a handle. It would do.

Someone put a pot of water on the fire to boil. Now all we needed was the filter, which was turning out to be more difficult than we expected. We'd all brought heavy woolen socks, not realizing that they'd need to double as coffee filters. Again, Michael gets credit for the solution.

"Paul, didn't you come straight from work? Maybe your dress socks would do the trick."

"Um, yeah, that might work," I said. He was right; I'd come straight from my job as an aide at a nursing home. "Problem is, I'm still wearing them."

Look, we were desperate. So I ran down to the lake, removed my socks, and washed them as best as I could in the clean, cold water. When I returned to the campsite, I placed the socks near the fire to dry. (My bare feet must have been frigid, but I was so wired that I don't remember feeling a thing.) I could tell my brother was excited. Like two mad scientists on the verge of a major breakthrough, we sat together on a tree stump and started to assemble our invention. Once the socks were dry, I pulled one through the hanger and fit it around the opening. Our friends looked on with skepticism, then flat-out disgust. One by one, they began saying thanks but no thanks. My brother and I giggled, knowing that our old man would be proud.

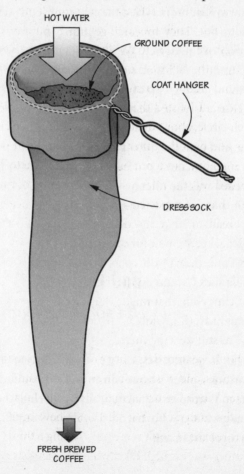

HOT WATER

GROUND COFFEE

COAT HANGER

DRESS SOCK

FRESH BREWED
COFFEE

I held our creation with pride. Michael and I couldn't have cared less about the dubious hygiene. We just wanted to see it work. The water boiled on the fire, and we scooped coffee grinds into the sock. As my brother gripped the handle and extended the sock over my cup, I slowly poured the water. The coffee dripped evenly into my cup, a rich deep brown, and that familiar aroma filled the air. Our friends started returning one by one. They appeared genuinely mesmerized by the perfection of the device. We all decided that I would have first-cup honors. As I took a sip, they looked on with anticipation.

"Now *that's* a cup of coffee," I said.

Then everyone else had a cup, and no one disagreed.

Paul Padial is a social worker in Manhattan. He grew up in the northern suburbs of New York City with his brother, sister, and anyone else who didn't have a home. His parents were always trying to fix things, especially people.

Bad Feelings

by Natasha Glasser

HERE IS SOMETHING I HOPE NEVER HAPPENS TO YOU: WHILE boarding a plane, I hope your travel companion never turns to you and whispers, "I have a really bad feeling about this flight." That's what my boyfriend, Jay, did on our flight back from Portugal last month.

Saying that to me is like offering an alcoholic a drink. After three years together, he knows I'm the Queen of Worry. Well, not the Queen, my mother still holds that title. But through some kind of nature/nurture double whammy, I'm the heiress to the throne. What really made the whole thing so disturbing, though, was that Jay just doesn't say things like that. He never complains that anything hurts, he never says he's worried about money or his job or the future of the earth . . . and somehow, even though I know how profoundly screwed the planet is, I believe him. That is the balance of power. That is why we work. That should not be monkeyed with.

I have a bad feeling about this flight. What do you even say to a thing like that?

I went with "Have you met me?!" And "What is the matter with you?"

"What? I have a bad feeling."

"And? You think we should get off the plane?"

"No. It's fine. Besides, it's too late now. Forget I said anything."

Right. Like that would be possible without a lobotomy. We got to our seats and saw we were in an exit row. Not a problem in and of itself, though a little more responsibility than I was comfortable with. But we had three seats to ourselves, so we weren't about to complain. Then I noticed that a woman and her small baby were seated directly in front of us.

"Could that be the bad feeling? Being trapped next to a loud, crying baby?" I was grasping, I knew.

He closed his eyes, gripped the armrests, and exhaled loudly. "It's not that kind of a feeling," he pronounced, like he was some kind of mystic. Then he took his little airplane pillow, put it on my shoulder and closed his eyes. "Just go to sleep," he mumbled. Amazing. He clearly didn't understand that worrying is an active pursuit.

I know enough about fear to know that it's all about lack of control, and you don't get a whole lot more out of control than hurtling through the sky in a massive soup can thirty-five thousand feet above ground. What I needed was the illusion of control. If I remained vigilant, maybe I'd, I don't know, see a piece fall off the wing and I'd be able to alert the pilot in a timely manner. Or maybe I'd be the only one to see my fellow passenger about to light his sneaker bomb while everyone else was sleeping or watching *The Tuxedo*, and I'd be able to alert the strapping young man sitting a few rows behind me to subdue him. (Whenever I board a plane, I always make note of a strapping young man just in case anyone needs to be subdued.)

I sat staring straight ahead at my upright tray table, annoyed that I'd let Jay get me so rattled. It would be fine.

I pulled out my *Us Weekly*. I always allow myself one trashy gossip rag per trip. They don't require much concentration, so an adequate level of vigilance can still be achieved. That occupied about an hour, and then it was back to tray-table staring. Several minutes went by without incident, and then we hit some turbulence. Nothing major, just the equivalent of a bumpy road, but it was enough to make me miserable. It did manage to wake Jay, though, so at least misery had some company.

"This is why you don't fly Air Portugal," I hissed. "I wanted to go on Continental." He had, in fact, found us a much cheaper fare on Air Portugal, and I had, in fact, been very happy about it. But that was then, and this was now, and now I was pissed.

"And how does that have anything to do with it?"

I was suddenly patriotic. "Well, *our* pilots know to go above or below the weather or something."

The bumping and lurching continued longer than it should have, I'm still convinced, but eventually things returned to normal and I relaxed a little. So did Jay. On my shoulder.

By this point, I was pretty exhausted, too. I wasn't going to sleep, mind you, I was just going to rest for a moment. I closed my eyes and put my head on Jay's shoulder. He reached up and stroked my hair. Some of the anger and fear went away. I remembered that he was kind. Nice. Good.

"Guess what, honey?" he mumbled. "I don't have a bad feeling anymore."

He was a moron. I was about to explain the profound karmic consequences of such a statement when the universe did it for me. As if on cue, the plane suddenly dropped like a stone and the lights started flashing and pinging, and there was a collective gasp as everyone clutched hands and hung in freefall.

Finally, after the longest five seconds of my life, we stabilized. Everyone exhaled and looked around with a shaky, smiley "That was a close one" expression on their faces. The plane stayed level. Which was good. But not enough. Because two seconds later it started to shake and shake—like the worst turbulence you've ever been in times ten. And it didn't stop. Then, one after another, the overhead bins opened and bags came tumbling into the aisle where the beverage cart was careening around wildly—unmanned since the flight crew had abandoned it to strap themselves into their little seats. I always look to the crew to gauge how bad any situation is. It always feels bad to me, but then I'll see that they are calm and chatting and having a diet cola and I'll feel better. I peered at them between the seats, and that's when I saw one of them make the sign of the cross.

They say you never know how you're going to react in an emergency. Introverted people take on gangs of muggers, weak people lift cars off babies—but I reacted pretty much the way I'd always figured I would; I just squeezed my eyes shut and tried to pretend none of it was happening. Jay, on the other hand . . .

I had his hand in a death grip, but he peeled it off. I opened my eyes and saw him undoing his seat belt. And standing up.

"What the hell are you doing?" I said.

"We need to get off," he answered. Very matter of fact.

"Are you insane? We're seven miles above the Atlantic Ocean. Sit down."

But he was already climbing over me, his eyes on the emergency exit to my left. I looked across the aisle, but the Portuguese couple sitting there had their blanket over their heads. Even worse candidates for an exit row than me. I glanced over at the flight crew again, but they were heatedly discussing something in Portuguese. An escape plan that they wouldn't be sharing with the rest of us, perhaps? An ejection capsule? Parachutes? I didn't know. What I did know was that my boyfriend was fiddling with the emergency door of the plane. He didn't really know what he was doing, but it also didn't look all that difficult to figure out.

Somebody needed to do something. I ran through my arsenal of skills: humor, reason, anger, pleading—all completely worthless at a time like this. Well, if I could not *be* the hero, at least I knew where he was sitting. Seat 39C. The strapping, Navy SEAL–esque guy I picked out earlier. There had been another contender I'd noticed during boarding, a muscle-bound guy with a barbed-wire neck tattoo, but he seemed potentially less selfless and wasn't seated on the aisle, I'd noted on my trip to the bathroom. Nope, 39C was my go-to guy.

I hesitated, wondering whether Jay could be trusted alone right now, even for thirty seconds. I studied him for a moment. He was still fiddling with the door handle, but in an idle way, like he was too paralyzed with fear to actually do anything that decisive. I unbuckled my seat belt and headed toward the back of the plane. The SEAL was just a few rows behind us, and everyone was too busy in their own panic spiral to pay me much mind as I headed back. I reached his seat and locked eyes with him. He looked relatively calm, all things considered, and was already undoing his seat belt even as I was saying what my face so clearly revealed: "I need you. Now."

He followed me up the aisle and we managed good speed despite the jerking and bucking of the plane. The flight crew was yelling something at us in Portuguese, but we ignored them. Jay and all the things that would be sucked out of the plane if the door were to be opened were still there, but his hand was on the handle. One good yank and bad, bad things would happen. I stopped a few inches from him. "Honey. Relax. You're fine. This isn't—"

There was a blur in my peripheral vision and my boyfriend went down. Sacked by the strapping Navy SEAL. And just like that, it was over.

The rest of the trip was bumpy, but bearable. Jay came to within a couple of minutes, and though it seemed pretty clear that the takedown had knocked some sense back into him, Navy SEAL guy and I still thought it best to keep him strapped into the seat between us for the remainder of the flight. Aside from a few mumbled "thanks" and "sorrys" and "I don't know what happened. I kind of lost its," Jay was silent. So I chatted with Navy SEAL guy (Dave) and learned that he was from Tennessee and had a fiancée named Tammy. Also, he was not a Navy SEAL. He had been in ROTC in college and now worked in pharmaceutical sales. All very light and small-talky, with no mention made of near-death experiences, mental breakdowns, or subduing.

When the landing announcement was made, he said good-bye to me and returned to 39C. I would've expected a parting exchange between the two men, but clearly some kind of macho wall had been broken. They instinctively understood that the only thing to say was nothing.

I also said nothing. I just took Jay's hand and held it. And when the plane finally touched down, and the passengers erupted in wild applause, we were whooping it up right along with them. It was a good feeling.

Natasha Glasser is a writer from New York City, and is currently working on a book of personal essays. She avoids flying whenever possible.

To Gravity!

by Joshuah Bearman

It was the turn of the millennium, gateway to the year 2000. The night when civilization would realize the promise Prince had made some two decades earlier on the title track of his fifth album. Or the night it would collapse for good: This New Year's was the one marred by the abbreviated terror, Y2K. Technology, we'd been warned, might fail us. The grid would go out. Planes would fall from the sky. All because some engineers forgot to add a couple digits to the nation's computer clocks.

My girlfriend, Ronni, and I live in Los Angeles, and we prepared for this cataclysm by renting a house in Palm Springs. Three friends, hand-chosen for their potential as survivors, joined us. Peter was mechanically minded; he had a theory that duct tape would become the most valuable commodity in a post-Y2K hellscape because it has such versatile applications. "Today's duct tape," he liked to say, "is tomorrow's gold." Dierdre, an artist, would perhaps create a new culture from the ashes of this one. Our childhood friend Tuesday is a waif and a lush and therefore offered no tactical advantage, but it seemed appropriate to have at least one wild card.

We arrived at the house on New Year's Eve day. It was big, filled with marble tile, linen closets, and a glass bauble chandelier in the foyer as big as a ship's wake. We stocked the vast pantry with survival provisions: a can of unsalted peanuts;

several bags of Flamin' Hot KC Masterpiece Cheetos; and a few bars of Scharffen Berger chocolate (82 percent cacao).

Out back was the pool and the adjoining Jacuzzi, an amenity that is perfectly suited to the desert winter, with its cold nights and clear skies. Whatever chaos lay ahead, we resolved to greet it with champagne in hand, floating blithely in the jet-froth whirlpool of that Jacuzzi. "Let 'er rip!" Ronni said, and flipped the switch on the heater.

Content with our mode of retreat, we whiled away the afternoon. Dierdre and Ronni made tribal masks from discarded beer boxes, sure to be of use in the New World, while Peter and I watched what we thought might be the world's last *People's Court* marathon. After a few hours, Tuesday wandered out back and discovered a problem: The Jacuzzi was not heating up. "We have to do something!" she shouted at no one in particular, before chugging a margarita from a coffee mug she'd found in the dishwasher. We gathered at the Jacuzzi's edge and stared at the bubbling, lukewarm water.

"We're screwed," someone lamented.

"I didn't touch it."

"Me neither."

"I saw Tuesday in there earlier. She probably broke it."

Tuesday downed her margarita and swayed indifferently. I noticed the text on her mug: I'M THE GRANDPA—AND YOU'RE THE DICKHEAD!

"Let's take a look at the controls," Peter said, snapping into action. He found the Jacuzzi's pipe and heating assembly behind a stand of birds-of-paradise, investigated, and pronounced it all sound. Then he lifted the drain cover to reveal a fistful of decomposing Cheetos floating in the well.

"Wasn't me," Tuesday volunteered without being accused. Half her fingers were stained with a deep orange dust.

But the waterlogged Cheetos were not the problem. The cause of our troubles, it turned out, was a fundamental imbalance. Like all built-ins, this Jacuzzi had a tile-topped membrane with a small trough for excess hot water to spill harmlessly over into the kidney-shaped pool. In our case, Peter observed, the osmosis was reversed: The waterline of the pool was too high, meaning the two bodies of water were joined. And the little heater's work was no match for the sheer volume of water in the pool, which probably measured about thirty feet long by ten feet wide.

"Looks like we have to figure out how to get about six inches of water out of the pool," Peter said. A quick calculation showed that this was, well, quite a lot of water. After an instinctive frenzy with buckets from the garage, we fell onto the grass, panting. Bailing was clearly not an option.

"Maybe we can use duct tape," Peter said weakly, realizing his theory was in tatters.

It was Tuesday who discovered the eventual solution when she tripped over a fifty-foot garden hose coiled up next to the house. As she got up and stumbled forward, brushing dirt from her legs and yelling at the hose for its insolence, Peter's face lit up. "Maybe a siphon would work?"

Yes! We would use the hose, in combination with a slight altitude differential across the backyard, to solve our problem. With gravity and atmospheric pressure on our side, the siphon would allow us to drain the water over an intermediate high point (the edge of the pool) without the aid of a pump. We lowered one end of the hose into the pool and surveyed the

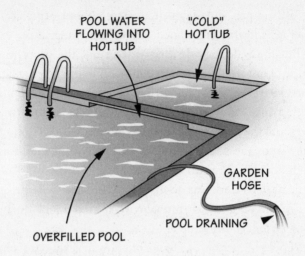

yard's topography for the best discharge position. Behind the garage was a ditch that ran to the curb, perfect for draining a half foot of pool water into the street.

Let me tell you, it takes tag-team sucking to get a fifty-foot siphon started. We almost gave up until, almost miraculously, a trickle dripped from the hose. Glorious water! We renewed our efforts, brought the siphon to full flow, and carefully arranged the hose for maximum gravitational pull.

Mt. San Jacinto was just starting to turn pink by the sunset when we left the physical laws of the cosmos to do their work. Around eleven-thirty, the Jacuzzi again became its own body of water, a discrete little lake that the heater quickly churned to a near-boil. Just in time, our survival plan restored, we piled in.

Poolside placards in hotels say that hot tubs and liquor don't mix, but I can assure you they mix very well in Palm Springs on the eve of destruction. We turned the TV so that we could see it from the Jacuzzi, where we soaked for hours, our fingers becoming too pruned to grip our margarita mugs. Only Tuesday managed to hold on to hers as the countdown came and went, followed by the world continuing to exist. No bedlam, no skeletons rising from the grave, only the steam floating quietly around us. Tuesday lifted her mug to make a toast. "To the future!"

Joshuah Bearman has many theories, most of which are based on little or no data. A contributor to the L.A. Weekly, McSweeney's, and The Believer, he recently compiled an entire volume of writing on the Yeti. He lives in Los Angeles.

The International Incident
by Tyler Cabot

THE FIRST THING YOU SHOULD KNOW IS THAT I'M PERPETUALLY late. The second is that I'm disorganized. Usually my method for getting out of a procrastination- or clutter-induced bind is a late-night scurry and/or an apologetic smile. But some problems are harder to fix. Like arriving at the airport less than two hours before your international flight leaves on one of the busiest travel days of the year—without your passport.

The day was December 24, 2004; the airport was JFK; the destination was Paris. It was 7:08 P.M., and my flight was scheduled to leave at 9:00 P.M. My older sister and I were on the way to the check-in counter when I realized my blunder: My passport was exactly where I'd left it the day before. On my bookshelf. In my studio apartment in New York City. A good sixty-minute cab ride (each way) from where I now sat. In other words, I was in trouble. Embarrassingly stupid, self-loathing trouble.

My first thought was, *What would Angus MacGyver do in this situation?* No, actually, that's not true. My first thought was to give up, head to the bar, drink away my shame, and catch a flight to Paris the next day. But being the older, responsible sibling that she is, my sister started quizzing me. Did I have any friends in town? Someone who could bring my passport, perhaps? No and no. Everyone had left town for the holidays. Plus, my spare set of keys were at the office.

Wait, no they weren't. Seth had them! I had given him my keys the night before. (He'd agreed to pass them on to an out-of-town friend of mine who was coming to New York in a few days and staying at my apartment while I was gone.) Unfortunately, though, after months of gastrointestinal unpleasantness (you don't want to know), Seth had undergone a colonoscopy a few hours earlier—he'd been sipping his chalky prep the night before when I'd stopped by, lamenting his sorry state. For all I knew, he was languishing in bed, still high on sedatives, watching *Alias* on DVD. It was a long shot, but I was desperate. I made the call.

Seth was not in bed. He wasn't even home. In fact, he was feeling much better. So much better that he was out eating Chinese food. (It seems the toughest part of a colonoscopy is preparing for it.) Solid citizen that he is, he immediately sprung into action. Here's what happened next:

7:13 P.M.: Seth leaves the restaurant in Chinatown, withdraws one hundred dollars from an ATM to cover cab fare, and heads to his apartment to pick up my keys.

7:20 P.M.: My sister checks in for the flight. I plead my case to the "travel counselor," begging her to hold my seat. She shrugs. It is Christmas Eve. It is a flight to Paris. Maybe.

7:32 P.M.: Seth reaches his apartment to get my keys. He tells the cab driver to wait downstairs.

7:40 P.M.: Seth reaches my apartment.

7:42 P.M.: A man waiting to check in overhears my phone conversation; assures me that traffic from Manhattan wasn't too bad.

7:46 P.M.: Seth leaves my apartment, passport in hand.

7:56 P.M.: I assure the travel counselor that my passport is on the way. I smile.

7:57 P.M.: I grovel.

8:00 P.M.: The flight is officially closed. The travel counselor won't give me a straight answer about whether I still have a seat.

8:10 P.M.: I begin pacing uncontrollably.

8:11 P.M.: I despise myself for being so forgetful. I start making false resolutions that I'm going to change my ways.

8:12 P.M.: A nice couple leaves the check-in area, wishing me luck, assuring me that everything will work out.

8:14 P.M.: Seth calls—he's here.

8:15 P.M.: I dash outside, grab the passport, and hug Seth as the cab driver offers a victory wave from the car. He had made it to JFK from the East Village in his fastest time ever: twenty-eight minutes.

With passport in hand, I dash to the counter and check in. I get my boarding pass. Deep breath, deep breath. Now to security. Breathe. On to the gate. I see my sister. She's waving and jumping up and down. Breathe, breathe. At last, a few dozen yards more, I'm there. And it's just the two of us. The last two people to board the flight. I give the flight attendant my boarding pass, and it's then that something really weird happens; something at first puzzling, and in the end, something miraculous (at least as far as international travel goes). The attendant tells us that our seat assignments have been changed—my sister and I have both been upgraded to first class.

Why did this happen? Perhaps it was a sympathetic gesture or an early Christmas gift. Maybe she simply liked my smile or had no other seats left on the plane. I'll never know for sure. But between sips of fine red Bordeaux—after I'd finished my chocolate *pots de crème*, but before falling asleep on my bed-size recliner—I reached two important conclusions: Sometimes being forgetful pays off. And the best MacGyver of all is a good friend.

Tyler Cabot, a member in good standing of Procrastinator's Anonymous, is a magazine editor in New York City. He arrives late to work every day.

A Spring in the Desert
by Wil S. Hylton

IT WAS ONE OF THOSE RADIANT NEW MEXICAN SUMMER DAYS, 105 degrees, with the heat oozing from the city pavement and the sun making pinpricks in the air, when Andrew said, "Let's go up in the mountains," and like Dorothy leaving Oz, I blinked and found myself alone, eight-thousand feet above sea level, trudging toward the Sandia Mountain ridgeline with an empty bottle of water.

Andrew was long gone—either far ahead or far behind; I had lost track and didn't care. We always hiked alone when we hiked together, meeting up at overlooks and trail intersections to compare notes. But that day, I had no notes. I had no thoughts. Four hours into the midday sun, thirst had taken over.

It was a maddening, desiccating kind of thirst, the kind that clots your blood and sucks the fluid from your eyes until your mouth is so dry you that you can hear the air swish in and out and your brain scrapes the inside of your head. I became aware of every speck of moisture in the landscape—the water content of a leaf, the blood inside of lizards. The sound of the breeze was a rushing river. The call of a distant bird was a squeaking, leaky pipe.

And then, suddenly, I had to piss. It came over me like a flood, this perfect storm of supply and demand. There was a spring in the wilderness, and it was me! I tore open my zipper with a dizzy grin and popped the cap of my water bottle, relishing the sound of it filling up. I knew there wasn't any real value to the fluid, that it wouldn't taste good, or rehydrate me, or replace any electrolytes, or even thin out the syrupy blood pumping through my veins, but I didn't care about any of that. I just wanted something to swallow. The only thing that mattered was that it was a liquid.

When the last drip topped off the bottle, I held it high and tossed it back, open-throating it gleefully. It was hot as bathwater, sulphuric and acidic, oozing across my tongue, but I promised myself that it wasn't disgusting. It was, I told myself, larger than it seemed, a metaphor for every important lesson I knew. It was the essence of self-sufficiency, of risks taken and answers found, options weighed and avenues gone down—this was what a man must sometimes do; what MacGyver would have done if he'd had to. High above the parched forgotten New Mexican treeline, a pint of Molson Golden.

Under normal circumstances, writer Wil S. Hylton gets his water from the spring behind his house in the Shenandoah Mountains of Virginia, which also, come to think of it, tastes a little pissy.

The Telltale Toilet

by Daniel James*

IN THE SPRING OF 2002, MY YOUNGER BROTHER, PAUL, GOT MARRIED in his adopted hometown of Reno, Nevada, and the whole family flew in to attend the wedding.

Let me tell you, Reno is *hot* in May. And after checking in to our hotel, a La Quinta Inn with a small swimming pool and free HBO, my wife and I discovered that our air conditioner wasn't working. We called the front desk, and a few minutes later a repairman arrived to investigate. He made a few calls on his two-way radio, then started poking around the electrical closet that housed the A/C. It only took a couple of minutes before he announced that he'd fixed the problem. We could feel the room beginning to cool before he even left.

Later that night, we returned to our now-cool room and went right to bed. Not that the temperature mattered by then. After the rehearsal dinner and an extended bout of merry-making back at the hotel bar with Paul and our older brother, Ted, I probably would have slept soundly on a bed of nails straddling the equator.

The next morning, my wife mentioned that she thought she'd heard muffled voices during the night. Strange, I thought. The TV hadn't been on. Maybe our neighbors were up late? It was a new hotel, so the walls were undoubtedly wafer-thin.

*A pseudonym.

As we were getting dressed to go meet my older brother, Ted, and his family for breakfast, I started to hear things, too: distant, unidentifiable voices. Upon closer examination, we agreed that the sounds appeared to be coming from the electrical closet. Could this have something to do with the A/C problem? Or maybe voices carried through the hotel's system of ducts and vents? We had no idea, but the voices were getting louder. We could only make out the occasional word; mostly it just sounded like gibberish. We decided to call the front desk again, which posed an interesting dilemma: We could only guess how the hotel staff would react to the news that their guests were hearing voices.

The clerk was very understanding, thankfully, and said she'd send the repairman back up. While we were waiting for him, the electrical closet issued the following announcement, the first complete sentence we were able to fully understand: "I've got a stopped-up toilet in Room 206, please respond."

Before my wife and I could even process this bewildering development, there was a knock on the door. The repairman. He walked in, went directly to the electrical closet, and returned with a two-way radio in his hand.

"I've been looking for this thing since yesterday," he said. "Mystery solved." He apologized profusely for any inconvenience he had caused.

My wife and I were having a good laugh over all this when something dawned on me. *Room 206? Isn't that Ted's room?*

In an instant, an idea began to percolate in some sadistic corner of my mind. Ted is three years older than I, and when we were growing up, he would torment me constantly. Sure, there was occasional violence, but mostly it was psychological warfare. His longest-running form of torture happened at bedtime. We shared a room, and every night, just after lights-out, Ted would initiate this brief conversation:

"Hey, Dan?"

"Yeah?"

"I *might* be coming over tonight."

Now, ninety-eight times out of a hundred, Ted would just close his eyes and go to sleep. But every once in a while he would silently pull back his covers, creep across the room, and pounce, nearly stopping my ten-year-old heart. That 2 percent chance was enough to keep my nerves on edge every single night.

And now it was payback time.

I walked across the hall and knocked on the door of Room 206. Ted opened it, and we exchanged pleasantries. Then I moved in for the kill. "Having a little bathroom problem, huh? You really have to be careful with these hotel toilets. They're not made for heavy-duty activity."

His reaction—or rather, nonreaction—was exactly what I'd hoped for. Ted was utterly dumbstruck, unable to even respond to what I'd just said.

"Yeah," I continued, "I was just down in the lobby getting a cup of coffee, and everyone behind the desk was talking about what a job the guy in 206 had done on his toilet. They said they'd never seen anything quite like it."

I could have milked this for the rest of the trip, but the look on his face led me to go easy on him. (Being the middle child, maybe I'm just not cut out for this kind of brotherly torment.) After about thirty seconds of letting this sink in, I explained the sequence of events surrounding the two-way radio, and when

I recounted the transmission regarding Room 206, a look of understanding and then embarrassment crossed his face.

And that's when I knew I had my revenge. That's when I realized that from this day forward (and for decades to come), this same look of embarrassment would cross his face each and every time anyone told the tale of the Telltale Toilet. It was an instant classic in the family archives, and I intended to tell it often.

Daniel James is a retired banker living in rural North Carolina. Now more than ever, he looks forward to Thanksgiving dinner at his brother Ted's house in Virginia.

MacGyver at the Bar
by David Wondrich

THERE ARE PEOPLE IN THIS WORLD WHO DON'T REGARD THE absence of a proper cocktail as a situation calling for immediate and extraordinary measures. I'm not one of them. Mind you, I don't require my every waking moment to be spent with fingers locked on the stem of a martini glass; there's a time and a place for everything. But when that time is now and I find my fingers closing on air, I start to get—well, not panicky, because that implies helplessness. "Motivated" is the word I want. So what if the vicinity holds no bar, no bartender, no cocktail trolley all stocked up with booze and the implements needed to beat it into the shape of a cocktail? As long as there's a drop of alcohol of (almost) any kind anywhere around, I *will* have that cocktail—even if it takes a little improvisation.

Luckily, when you strip off all the flair and silliness with which it's so often invested, mixing good drinks is a pretty simple business. You don't need to know five hundred recipes. You don't need to write about booze for a living, as I do. You don't need jiggers, speed-pourers, neon liqueurs, or birdbath-size crystal martini glasses. All you need is a little imagination, a little knowledge of the basic principles of mixology (measure everything, make the drink as cold as possible, and don't overdo it with the sweet stuff), and, above all, the will to get it done. As proof, I offer three examples, occasions when I was far from home and MacGyvered myself (and various other interested parties) a reasonable cocktail against very long odds.

My first success with field mixology came in 1983, when I was twenty-two and found myself stuck in a cheapo motel in Greensboro, North Carolina, with three other guys and nothing to drink. (It's okay, we were a band.) It was the last night of our tour and we were all down to pocket change. Half an hour's scrounging through backpacks, drum cases, seat cushions, and whatnot yielded just enough money to pop around to the state store and pick up a pint of their cheapest vodka. With twelve cents left over, mixers were out of the question—as was ice (the motel charged a buck for it). Even a bunch of broke rock-and-roll musicians could see that a couple swallows of warm Popov are almost as bad as no drink at all.

Fortunately, there was a cafeteria-style barbecue joint next to the motel. That gave me the idea. I would have to be quick—and discreet—but the prize would cost nothing more than a dirty look from the cashier.

Keeping my head low, I slipped through the door, found the drinks station, grabbed one of their massive Styrofoam iced-tea cups, filled it with ice, snapped a cover on it, pocketed three or four lemon wedges, and slipped out again—all in less than sixty seconds. Three minutes later, we were back in our room. I peeled the lemon wedges with a Swiss Army knife, poured the vodka into the cup, shook the bejeezus out of it, punched a thicket of holes in the lid with the awl (that pointy thing on the back of the pocketknife), strained the contents into four motel glasses that had been cooling under the faucet (better than nothing) and twisted a nice, fat strip of lemon peel over each. A minute later, A Blind Dog Stares, as we were calling ourselves (yes, I know it's a stupid name), was sipping bone-dry vodka martinis. Not too shabby, considering.

In the two decades that followed, there was a good deal of competent—if I do say so myself—but essentially unexciting fieldwork that I'll skip over. I learned a lot about making do. (Did you know that a beat-up old coffee mug chilled in the freezer for half an hour will keep your drink cold far longer than any fine-crystal martini glass in existence? Or that you can make a fairly palatable margarita with tequila, lime juice, and Mrs. Butterworth's?). One thing that I could never work around, though, is the absence of vermouth when a martini jones strikes. (While I'm fine without it in a vodka martini, with gin it's a different story.) So when, a couple of years ago, my wife and I found ourselves all alone at a friend's country house in upstate New York, stocked with all the good things in life—bottle of Tanqueray, real martini glasses, antique cocktail shaker, lemons, lots of ice—save that one, we came close to despair. I mean, it's not that we *needed* that martini, mind you; it's just that it had been a long drive and there was a patio and a sunset looming and . . . okay, we needed it. But there was no way we were going to get back in the car. An insolvable problem.

Then the inspiration: What is vermouth but oxidized, lightly fortified wine that's had herbs and spices steeped in it? And what was that behind the house but a large and well-tended herb garden? A rummage through the refrigerator turned up half a bottle of Pinot Grigio. God knows how long it had been open, but in this case it scarcely mattered. We were in business. A sprig of rosemary, a little thyme, some chamomile flowers— I had no formula, but those sounded about right. Perhaps a strip of orange peel for aroma and a couple of sage leaves for their bitterness.

Now, a big part of the vermouth-making process is aging: It takes months for the herbs to impart their flavor to the wine and for the wine to oxidize. If we had had to wait that long for a martini we would surely have died. But there is such a thing as a microwave, and this house had one. Quick and dirty, but effective. After a couple minutes of shaking everything together in the bottle, I decanted it into a glass bowl and nuked it for half a minute or so. Result: crude, but adequate. Martinis ensued.

My third and most recent booze-related MacGyverism happened in the summer of '05. Once again, I found myself in need of ice, only this time I was in France with some fellow journalists, touring Cognac distilleries, as pleasant a form of work as has ever been devised. There was a slot in our itinerary for an afternoon boat ride down the Charentes, the sleepy little river that runs through the Cognac region. What could be more enjoyable, we figured, than to spend a June afternoon floating downstream, shooting the breeze, and sipping cocktails? We came prepared. I packed a cocktail shaker, a strainer, and a juicer. Someone else brought some bitters, someone else some Cognac, etc. All we needed were the cups and the ice. A quick stop at the local hyper-market (as they call them) and we'd be all set. The cups were no problem. But who knew French supermarkets don't sell ice? What kind of a country is this? The boat left in ten minutes; there was no time to go anywhere else.

Just as despair was setting in, it came to me in a flash— they've got a fish counter, right? Our best French speaker was dispatched. Five minutes later, he was back, a big bag of ice chips in each hand. The cocktails only tasted a little bit like fish.

David Wondrich is one of the leading authorities on the cocktail and its history. He lives in Brooklyn in a house full of bar gear.

LOVE, LUST, AND OTHER COMPLICATIONS

I RECEIVED HUNDREDS OF SUBMISSIONS FOR THIS BOOK, AND I'D SAY at least 40 percent were about love, like, trust, lust, dating, hating, breaking up, or cheating.

Initially, this surprised me. *Romance?* Granted, Mac wooed at least nine women in season one alone, but his love interests always seemed unnecessary—no one watched *MacGyver* for the mushy stuff. This book, I'd always imagined, would be about tangible solutions to concrete problems. Blown gaskets. Leaky faucets. Severed bypass lines. What's love got to do with it?

I was profoundly wrong. And the more I thought about it, the more I realized that I should have known better. The one thing all good MacGyver stories have in common is high stakes. Finding a way out of the situation—it's gotta *matter*. For Angus MacGyver, that was never an issue. The man saved the world once a week. But the stories in this book are by (and about) real people, not secret agents on TV. And real people don't often find themselves in a situation that requires them to save a life, much less all of Western civilization. Except for doctors, stuntmen, and pro poker players on ESPN, most real people never face higher stakes than the ones they face in love or its immediate precursor, deep like.

Just ask Susan Burton of Brooklyn, who tried to write her way through a breakup (and not in the way you might expect). Or Robin Romm of Berkeley, who exploited her new boyfriend's ego to get him to wear better pants, saving their budding relationship in the process.

As these stories rolled in, a pattern emerged. More so than in any other chapter, the protagonists of these adventures in romance didn't merely need to be clever—they needed to be clever *fast*. In his story, Chuck Klosterman talked himself into what seemed like an inescapable corner on Valentine's Day before somehow wiggling free at the very last moment. Tiffany Funk, an American college student studying abroad, had less than ninety seconds to perform emergency surgery on her skirt before her date arrived. And Vince, the brilliant antihero of Francine Maroukian's excellent story of high-wire two-timing, well, let's just say that Vince gets the Save of This Book Award.

It's a cliché that in moments of extreme desperation, human beings are able to perform beyond their abilities. But when it comes to the most important moments of all—the moments that make or break the bonds of love—it's a cliché because it's true. The MacGyverisms in this chapter aren't merely acts of improvisational genius; they're potentially life-changing acts of improvisational genius. And in these situations, duct tape is of limited value.

We begin, naturally, on February 14.

The Valentine's Day Miracle

by Chuck Klosterman

THIS IS A STORY THAT MAKES ME SEEM CLEVER, BUT IT ACTUALLY JUST proves I'm stupid.

When I was twenty years old, I had no idea that adults celebrated Valentine's Day. It seemed like something only children would do, and I assumed normal people ignored this holiday completely. But (at the time) I was dating a very attractive woman who was twenty-nine, and she saw this situation differently. On the afternoon of February 14, we went to see *Groundhog Day* (this was 1993), and then she gave me all these very specific, very thoughtful gifts: leather gloves (which I needed), the children's book *Go, Dog, Go!* (which I had casually mentioned liking when I was five), and the Soul Asylum album *Grave Dancers Union* (which further proves that it was 1993).

I didn't get her anything.

The idea of buying a Valentine's gift had never even occurred me; it seemed like something from the 1950s. But as my girlfriend handed over each of these inexpensive (yet deeply personal) presents, I could feel my blood becoming slowly infused with liquid dread. I could not believe I had managed to exist for two decades without knowing that women care about Valentine's Day. And I started to get extremely nervous, because I did not know the degree to which I would be penalized for my ignorance. What if this was the kind of thing that caused people to break up? I mean, if I didn't even know that it was normal to give people presents on Valentine's Day, who knows what else I was confused about?

Seeing no other option, I started lying.

I thanked this woman for the gifts, and then I began to explain my Byzantine criteria for selecting her particular gift. "I didn't just want to buy you something boring," I said. "I wanted to give you something more creative. I wanted my gift to somehow directly reflect the context of our relationship." None of these words had any meaning. I was just throwing out vague generalities; it was like I was the press secretary for a president who had accidentally declared war on Canada. "I wanted to give you something that would truly represent the spirit of the occasion in a less-than-orthodox manner," I said.

I began walking toward my closet, building suspense the entire time; I was giving every indication that I was stashing something pretty awesome behind its wooden door. *But there was nothing in this closet except my clothes.* I had no idea what I was doing. I was essentially waiting for an object to spontaneously build (and gift-wrap) itself. "I really hope you like this," I continued. "I hope you don't think this gift is stupid." I began opening the closet door without any plan and without any hope. There was no future.

I looked inside.

The closet was filled with clothes and darkness. My clothes. My darkness.

And then I instantly became a genius.

"You know, when I was in high school, I never had a girl-friend," I said, which probably explains why I never knew people cared about Valentine's Day. "But all my high-school buddies did. And they would always give their high-school football jackets to their girlfriends, and I was very jealous of that exchange. It made me feel very incomplete and alienated. So I was just wondering if you would want to wear my high-school football jacket." This entire scheme hit me like a bolt of lightning. It was completely extemporaneous and weirdly detailed, and I have no idea what it even meant (or what it truly says about me).

My girlfriend began to cry. Then she said that would make her very, very happy. So I gave her my high-school football jacket. Which I had not worn in three years. And which I actually had two of. I had one jacket from my sophomore year, and one jacket from my senior year. (I gave her the newer one.) And there had never been a moment in all of high school when I had wanted to give either of them to a girl, because that always struck me as crazy. But now everyone was happy, albeit for very different reasons.

I am not proud of this story. But it happened.

Chuck Klosterman is a columnist for Esquire *and is the author of* Fargo Rock City, Sex Drugs and Cocoa Puffs, *and* Killing Yourself to Live. *He still has the other jacket.*

The Golden Earrings

by Tiffany Funk

As a junior in college, I studied abroad in Madrid, Spain, and whenever I had the chance, I traveled alone around Europe, sleeping in hostels and meeting people from all over the world. One November weekend in Paris, I met a (very cute) rugby player from New Zealand in the bar of the hostel where we were both staying. We had a few drinks and talked late into the night. (Yes, that's all that happened.) On Sunday afternoon, before I left for the airport, I gave him my contact information back in Madrid. I was sure I'd never hear from him again.

So you can imagine my surprise when, two months later, I got a call from you-know-who. He was in Madrid for a day on business, and he wanted to spend the evening together before he caught his flight back to Brussels, where he was living at the time. Flattered but flustered, I told him that I would be more than happy to show him around a bit and take him to one of my favorite Mediterranean restaurants. He told me that his hotel was only a few blocks from my apartment, and said that he'd be there in a few minutes. I hung up.

A few minutes? I ran to the bathroom and dumped out my cosmetics bag. Lather, rinse. Foundation, blush, eyeliner, eye shadow, mascara, lipstick. After spending way too much time applying my makeup, I kicked off my ragged jeans and pulled on this soft wool skirt I had just bought at Benetton. I'm generally a T-shirt-and-jeans kind of girl, and this was the most expensive article of clothing I'd ever owned. Now seemed as good a time as any to break it out.

Before I could zip up the skirt, I heard the door buzz. As I buzzed the rugby player in, I yanked on the zipper. It was stuck. I panicked. I yanked—hard—and the pull tab broke off in my hand. I stared at the tiny piece of metal, stunned. I was on the sixth floor, which meant I had less than ninety seconds before he'd be at the door.

What to do? As a traveler, I had limited resources. I had no safety pins, and my travel sewing kit was useless considering that I now had less than a minute to finagle my way out of this situation. I cast my eyes around the small apartment, looking for a clue. And that's when I caught my reflection in the mirror. My earrings! They were small gold hoops, and they just might work. I pulled them off and stuck them through the fabric, pinning the halves of the skirt together at the top and middle of the zipper. Just as I finished untucking my shirt and smoothing it out, he knocked. I took a deep breath, opened the door, and smiled widely. My kiwi friend was none the wiser at my apartment, at the restaurant, and at the fountain at Plaza de España, where we said adios.

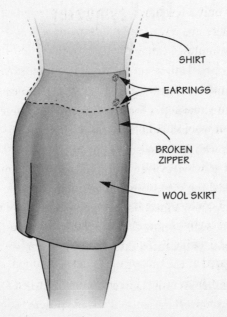

SHIRT

EARRINGS

BROKEN
ZIPPER

WOOL SKIRT

I learned three things from this experience: 1) The more expensive the clothing, the more the hassle; 2) Always keep either a pair of hoop earrings or safety pins on you wherever you go; and 3) If you can't wear your favorite ripped jeans with a guy and be comfortable, he's just not worth it. Even if he has an accent to die for.

Tiffany Funk is a Wisconsinite by birth, an Illinoisan by choice, and a humanities instructor and artist by trade. She lives in Chicago with her cat, Lex Luthor, and her American jeans-and-T-shirt kind of guy, Jeff.

The Keymaster

by Francine Maroukian

I DIDN'T THINK OF IT AS SLEEPING AROUND; I PREFERRED "ENSEMBLE dating." But anyway you look at it, there were three of us that summer: me, my old should-we-try-again boyfriend Lawrence, and Vince, who was too young for anything but bed. At the time, both guys were New York catering waiter/actors who often worked the same jobs and went on the same auditions. So Lawrence knew Vince. But he didn't know *about* Vince. And he certainly didn't know that Vince carried around my only spare set of keys because I didn't like to get up late at night to buzz him in.

One summer evening, Lawrence and I returned to my apartment after a concert in Central Park and ordered some Chinese food. A few minutes later, someone started noisily fumbling around at my apartment door. There were a few moments of hurried confusion—how did the delivery guy get in without buzzing, where's my wallet, I'll get the plates—before the door pushed open and there stood Vince, wrist still twisted from turning the keys in the lock.

I'm not sure how long it took—it felt like forever—but in one easy gesture, Vince untwisted his wrist and withdrew the keys, catching them in his fist. Then he strode across the room, extended his arm and said, "Hey, do you know you left your keys in the door?" He dropped the keys in my outstretched hand and I instinctively curled my fingers around them, preventing Lawrence from seeing that the key ring wasn't mine. Vince shifted his shoulder-slung backpack to the other side and continued. He was working on a scene in the neighborhood and thought he'd stop by. My neighbor let him in and how were we doing. It was all so brilliantly natural and friendly. That's when I knew Vince had the heart of a snake and a genius for cheating. I'm not ashamed to say that I admired him. The cheater in me still does.

Francine Maroukian, author of Esquire Eats, Town & Country's Elegant Entertaining, *and* Chef's Secrets, *is a former New York City caterer who never hired a waiter she didn't love. At least, not twice.*

Bad Pants

by Robin Romm

IT WAS A REAL PROBLEM, DESPITE WHAT YOU MIGHT THINK. A NEW relationship is a delicate thing, and the pants are important.

My friend Mira has a lot of requirements. She can't stand guys who carry man-purses. She has specific standards for the shape of a man's shoe (can't be too pointy, no noticeable heels). I knew she'd understand the gravity of this. I called her after the third date.

"It's just—he has really bad pants."

"Oh no," Mira said. "What kind of bad?"

"For one thing, the legs taper—they're like big funnels." I told her how big they were, how every pair would forgive an extra hundred pounds. I explained the garish patch of purple fabric sewn crudely to the butt of a pair of brown corduroys. I mentioned the gaping hole in the knee of his favorite jeans—which, by the way, weren't even real denim.

"Oh God," Mira said. "They're that fake denim? The pale blue cotton kind?"

I don't consider myself shallow. I was raised in a hippie town in Oregon. During my adult life, I have tried meditation, yoga, therapy, vegetarianism, and civil rights investigating. But the pants thing—that was deep, too. I liked Matthew a lot—I'd felt an instant attraction to him during the first class of the fiction workshop where we'd met. But here we were, living in a city, trying to make beautiful things together, and those pants were pulling him down. He wanted to eat crepes in those pants. He wanted to go get beers in those pants. How could he write sentences like the ones he brought to our writing workshop and still participate in this kind of aesthetic slaughter?

So, I would try to distract myself with his white teeth, the green flecks in his eyes. He was a sensitive man, and we'd already had a few arguments about my criticisms of life and people. Once I commented on another classmate's body odor and Matthew rolled his eyes. He was pro–unconditional love and nonjudgmental friendships. I had to be careful.

For about three weeks, we focused on not wearing clothes at all. This worked well. We spent a lot of time in his basement apartment, pantless. But eventually, things progressed. There were restaurants to visit. There were movies to see.

Then I went to visit my parents in Eugene for the weekend. My father, in his retirement, was learning how to cook. My mom sat at the kitchen table with a glass of iced tea while my dad tried to cut potatoes with a butter knife. Sweat beaded across his upper lip. He assessed the potato, positioned the dull knife, sawed it back and forth. I wanted to tell him to use a real knife. I wanted to show him how to hold it correctly.

"I never knew your father was such an excellent cook!" my mom said. "He's turning over a new leaf." My mother sat there, instructing him: "You put the fish in the square glass pan and then you drizzle olive oil, like that." He opened the olive oil, dropping the cap behind the oven vent. As he went to retrieve it, the oil slipped and he spilled several cups on the fish.

"Good. Oh my! I've never seen the fish look so perfect in the dish! Look at the way you drizzled that oil! You're an olive oil natural!" Her brown eyes gleamed with praise. And my dad got lighter and lighter. He was all puffed up about his olive oil and potato capacity. He was a wonderman. A miracle. A man who could slice and drizzle! For twenty-seven years, my dad had walked through that kitchen as if it were a place of extreme mystery. But for the three days I stayed with them, he couldn't wait to get in there. He made scrambled eggs. ("So fluffy!" my mom said.) He made turkey sandwiches. ("What tender meat!") He made burgers. ("Oh, you are such a pro on the grill!")

On Sunday night, Matthew picked me up from the airport. He was wearing his brown cords. We chatted about my visit and held hands as we drove through the city. We got back to his underground apartment and started kissing.

"You know," I said, "you've got a really sexy body." He paused, waiting for more. "Most people don't have the body you do." It was working. He was puffing up, holding out his chest. "You should wear tighter pants to show it off." He glanced down at his pants.

"You think?" he said. He went over to his mirror and turned sideways.

"I'll help you find some," I said. It came out overeager but he didn't seem to notice. He was too busy floating around. He was a man with a hot body. Absolutely, the pants would come.

Robin Romm lives in Berkeley, California, where she continues to fight the good fight. (She recently helped Matthew buy two pairs of flat-front chinos.) Her stories have appeared in Tin House, One Story, Threepenny Review, *and other magazines.*

A Girl's Guide to Mending the Unmendable

by Susan Burton

THE SUMMER I GRADUATED FROM HIGH SCHOOL, I HAD A BOY-friend who now, years later, I think of as a creep. Since he had my husband's name, I won't call him that. Instead I'll call him what I called him later, which is Shane.

Shane was my friend Lisa's older brother. I was seventeen and he was twenty-three. Shane was the only person I'd ever been friends with who'd been alive in the 1960s. Of course, he'd only been a toddler, but that wasn't the way I thought about it. Shane was so old that he'd been alive in history.

He was also the first person I knew who'd been engaged. The girl he'd been engaged to was named Kathy. They'd lived together during college, but then she'd run off with the man-ager of Wild Oats, a local health-food store. I worked at a health-food store, too. But my health-food store was called Al-falfa's and it was the type of place where there was less hum-mus and more pâté. There were a lot of Colorado moms in Volvos. I would watch them as they wrote checks. They had big diamond engagement rings that they now had to figure out how to coordinate with the turquoise jewelry they'd dis-covered when they moved to the West later in life.

One morning that summer I looked up and Shane was in my line. It was the first time we'd been together, by ourselves, with-out his sister. As usual, he looked pale—he was more of a reader than a river-rafter, and was always recommending books. Today he was holding a single can of all-natural dog food. "I ran out of food for the dogs," he said. Instead of putting the can on the counter, like most people did, he put it right into my hand.

And then it started. Shane was the best boyfriend I'd ever had. I felt so grown up, walking across his lawn wearing lipstick and cowboy boots. Other nights I'd ride my bike home from Alfalfa's and he'd be there, waiting for me outside my house. I'd get in the car and we'd start driving and he'd look over at me in the passenger seat and tell me the thing I was doing with my thumbs was a known sign of genius. He thought I would be the one in my class to become famous first.

Soon it was time for me to go to college, and for Shane to go back to his job teaching at a junior high. Before I left, Shane wrote me a poem and gave me a silver barrette from one of the Native American jewelry stores. Flying across the plains in August, fingering the barrette in my hair, I decided I would miss Shane, but not in a terminal, serious, sad-girl way. It had been a perfect summer.

Then I was at school in Connecticut. It was humid, even in the middle of the night, when we'd all be crammed into a booth somewhere, singing loudly to the Eagles. Classes were slow to start and everything was still fun. It was 1991, the last year before e-mail really took off, and everyone still wrote letters. I got good letters. Once, Shane wrote, "As soon as I sealed this, I felt like writing more." And another time he wrote, "I love you very much." I thought about this a lot. We'd never said it to each other in person. But now, I wondered.

We hadn't planned on staying together, but instead of thinking of Shane less, I was thinking of him more. Unfortunately, his letters decreased in direct proportion to my thoughts.

Worse, in recent notes, he had begun to mention a first-grade teacher named Tina.

One night in March, I brought the cordless phone up to my top bunk and I called him. He said, "Well, I have some good news. I'm getting married. To Tina, the first-grade teacher."

And that was when I knew: I loved Shane.

I wrote reams of five-page papers, took finals, and returned to Mountain Time. Back home in Boulder for the summer, I couldn't get Shane out of my head. Wearing a skirt, I would ride my muddy mountain bike fast up Arapahoe Road, which I knew he had to drive on a lot, hoping he would see me. I thought the skirt plus the mud made me look wild and pretty at the same time. At night, I'd tilt my halogen bike light upward so that it would cast a mysterious glow upon my face. I'd lean forward, pedaling furiously. *Drive by, Shane*, I'd think. *Drive by and see me.*

Sometimes in this fantasy, it was Tina who would notice. "Look at that girl on her bike," she'd say. "Look at how fast she's going. And she's so pretty." And Shane would glance away from the road, and it would be me. There would be a silence as he turned back to the wheel, him deciding. They'd slow down at the light. He'd get in the left lane, turn on his blinker. "That's Susan," he'd finally say. "That's Susan?" Tina would say, whipping around just in time to catch me flying over a hill, my hair the lightest thing in the dark night. And then they would drive back to his parents' house, Shane filled with regret, Tina feeling inadequate, and the next day, while Tina was out, he'd be overcome; he would come to buy a can of dog food at Alfalfa's.

Of course that's all it was: a fantasy. One day I resolved to get over Shane before the wedding. It seemed important not to think about him once he was married. Pathetic, but wrong, too. The problem was that I had no idea how to go about it.

The one bright spot in my summer was my fiction-writing workshop. I'd signed up for a class that met two evenings a week at the University of Colorado.

One evening I rode home thinking about the story I'd written for that night's class, a story I called "Sunday Morning." "Sunday Morning" was told in the voice of a twelve-year-old girl whose parents were on the verge of divorce. Like every story I'd ever written, it was based on a real-life experience.

In one part of the story, I said that the house was always cold because the father didn't like to waste money on the heating bill. In another part I mentioned a photograph of girls in matching dresses at the country club. My classmates thought the country club symbolized an earlier, golden era, and the cold house meant that now the money was gone and soon the family would be, too. Now, swooping down the bike path, I considered whether, in addition to not getting along, my parents had been losing money. It wasn't a point I had intended to make, but it was right there in the details for the class to discover. They could see things about my own life that I couldn't.

And that's when I got the idea. I would write the exact story of my relationship with Shane. I would introduce the class to me, the static protagonist, and they would tell me how to act. They would not only tell me how to get over Shane, they would tell me why it had all happened and what it all meant.

The next morning, during my shift at Alfalfa's, I sipped from a giant glass of hot tea with soy milk and planned out the story. The action would take place on the day of the wedding. It would be a race-against-the-clock scenario: The protagonist would have twenty-four hours to get over her ex-boyfriend. This struck me as an elegant conceit, and also a useful one. If I woke up on the day of the wedding and wasn't yet over Shane, I wouldn't have to sweat. Thanks to the class, I'd have a twenty-four-hour cure.

That evening, I sat at the computer. "On the day of the wedding," I wrote, "I worked a double shift. In fourteen hours, I made thirty-nine goat-cheese-and-basil sandwiches, wrapped 106 lunch boats in tight Saran Wrap, and sliced through sixteen different ten-inch cakes in the course of waiting on at least 357 customers. I was counting to keep my mind off of some other things." I sat back. *There*, I thought, satisfied. As long as I switch myself from cashier to deli worker, nobody will know this is me.

I called the story "Waiting" and my character "Katy." Like Katy behind the deli counter, I would be stationed in the classroom, waiting to be delivered a satisfying end.

When the evening came I pedaled to class at race-day speed, full of anticipation.

The class was a mix of undergrads and people a few years out of school—like a guy who rode the free trolley from coffeehouse to coffeehouse all day, and an aerobics instructor with tan, very shiny legs. Then there was our teacher, George.

"Let's start with 'Waiting,'" George said as we settled in.

This was it. I readied myself for the wise and luminous insights of my classmates.

"Okay, just starting with the first paragraph?" said the aerobics instructor. "I highly doubt someone would count everything like that. And I don't think it's very realistic that somebody could remember all that in their head."

"Maybe she was writing it down as she went along," somebody suggested. There were nods. I returned the nod. I leaned over the paper and made my first note.

"Counting," I wrote, followed by a question mark. *Small potatoes*, I thought. *Bring it on! What else?*

"I was wondering where she went off to college," someone said. "I mean, it's back East, but where?" Murmurs of agreement.

This was not going according to plan. *Come on: How should she get over him?!*

"Wait, wait, wait," said the aerobics instructor. "About the bride. Or, actually, the wife. I thought it was weird how Katy never even thought about the wife. You know? She never even wondered what she was like."

I had to hand it to the aerobics instructor. Sometimes she would come out with something really sharp. "Why didn't she think about the wife?" I wrote, with a big, filled-in arrow next to the note. *Why* didn't *I ever think about Tina?* I wondered.

"I guess, also, I didn't really have a sense of Shane," someone added. "I mean, it says he's a teacher, and stuff like that, but I mean more about why he got married so quick."

"Why Shane got married so quick," I noted. I was starting to see that there might be a problem. The class had some of the same questions about the characters that I did.

"Let's talk about Katy's feelings for Shane," the teacher said. *Okay, right idea*, I thought.

"There was this one part that I thought was pretty telling," said one of the older students. "Where she says on page three, midway down the page"— there was the sound of people flipping— "I don't think I loved Shane until he told me he was getting married."

"Exactly," George nodded. "But I think it's getting buried in there. You might want to make it bigger," he said, addressing me directly. "Or earlier."

He tipped back in his chair. "What are Katy's true feelings? Is this a story about social status? Or is this a story about true love?"

I didn't like the way the question was framed.

"Social," the aerobics instructor said. "Definitely social. I mean, if they'd been in love, it would have clicked when they were still together."

No! This was not it. As everyone bobbed their heads in agreement that Shane and I hadn't been in love, I looked very intently at my paper. "Why didn't anything click," I wrote. Then I couldn't write the rest of it so I just made a long dash.

"We don't know much about Katy," said George. "She went off to college, she enjoys mountain biking. But aside from that: Who is she?"

The class was silent. Then George turned his entire body and looked me straight in the eye. *Oh, God,* I thought. *They know.* "Susan, you might need to do a little prewriting about Katy. Get a sense of who she is, figure her out a little more." Yes. I nodded vigorously.

"Okay, well, we should really move on to the next one," said George. "Susan, I guess my final comment would be that this story needs an end. But it develops well. Nice job."

I sat there blankly. *Story needs an end?* I thought. *No duh it needs an end! The story needing an end is the whole reason we are here!*

And then we just started talking about some other person's work. I sat very still and looked around the bright room, all of us at the desks in a big circle. Nothing had happened. I was still the same. It made me feel lonely, the way they'd talked about my life.

But as the days went by, I felt better. The class had acted like my story wasn't so unusual. Just some old thing with an ex-boyfriend. It made the whole thing feel smaller, like less of an emergency. Dozens of girls now riding their mountain bikes through these very foothills had dated boys and been hurt. It happened all the time. It was a known thing and it made people sad.

Now I rode my bike along Arapahoe Road at the speed of a regular girl, wearing regular clothes, like shorts, and pointing my bike light in the regular direction, straight ahead.

For a while I still thought about Shane, and then, suddenly, I didn't think about him at all, except on every birthday, when I would think, *I'm still not as old as he was.* And then I was twenty-three, and that was it: I didn't think about him anymore.

Susan Burton is a contributing editor of the radio program This American Life and the coauthor of Come Back to Afghanistan: A California Teenager's Story (Bloomsbury, 2005). She lives in Brooklyn with her husband and their son, who is by far the most ingenious member of the family.

Here Comes Security

by Kevin Fedarko

IN OCTOBER 1996, ON THE NIGHT BEFORE MY FRIEND BRETT MOUTON got married in New Orleans, he presented each of his grooms-men with the gift of a Leatherman all-in-one tool. It was a pretty swanky model, with all the bells and whistles: regular pliers, needle-nose pliers, wire cutters, hard-wire cutters, knife, scissors, wood/metal file, large screwdriver, ruler, bottle/can opener, wire stripper, the works.

As groomsmen's gifts go, the Leatherman was a pretty cool idea. In addition to its MacGyverish cachet and flat-out use-fulness, the tool also symbolized the kind of seat-of-your-pants sensibility that had led Brett to his fiancée in the first place. But that's another story for another time.

Teetering on the threshold of marriage, Brett had begun to wonder whether he'd made the right decision. The funny thing is, *I* knew from the start he was marrying the right girl. His fiancée was a willowy blonde from Long Island named Elizabeth Kelly. Elizabeth was elegant and incredibly funny, and her eyes harbored an elusive shade of blue that mirrored the cobalt mysteries lurking deep beneath the lagoons of the South Pacific (or so Mouton had insisted to me, in language almost that overwrought, after he'd met her for the first time three years earlier). In short, Elizabeth was far better than any-thing Mouton even remotely deserved.

Unfortunately, *he* would only know this after the ceremony

was over. Like any man about to commit the rest of his life to another human being, he was a wee bit anxious. "What will it be like to never feel the soft lips of a strange woman . . . ever again?" he'd asked me earlier that day. At the moment, he was a bundle of nervous energy as he began to greet people at the entrance of a private room in Antoine's, a storied restaurant in the heart of the French Quarter, where the rehearsal dinner was about to begin.

It was here that I witnessed his first fatal mistake. Brett was so excited meeting and greeting friends and family from far away and long ago that, despite the platters of jambalaya, rice and beans, and barbecued shrimp on every table, he barely ate a bite. But he seemed to love the taste of Abita beer, the local brew—in the soft glow of the electric candelabras, I watched my friend pound beer after beer after beer.

Brett is a bonafide Cajun. He is small and he is muscular and he can do two things well: eat crawfish and drink beer. (His favorite joke that night: Q: Do you know why Cajuns don't eat M&M's? A: They're too hard to shell.) On this night, though, it was as if he were a human sponge. His cheeks grew rosy. He smiled a wide grin. And he didn't talk much. But here's the rub: By midnight, he still didn't really look *drunk*.

As the rehearsal dinner was winding down, Brett suggested those of us still hanging around head across the street for a nightcap, to a bar he liked called the Old Absynthe House, on Bourbon Street. Not wanting to abandon the guest of honor, a small congregation of family and friends (myself included) accompanied Brett to the bar, where we each began treating Brett to drink after drink.

This is when Brett broke another cardinal rule of drinking: He switched to hard liquor in the form of mixed drinks. At one point, he actually drank an entire Sex on the Beach shooter, which, for those of you unaware, consists of melon liqueur, raspberry liqueur, cranberry juice, and pineapple juice—oh, and lots of vodka.

Which brings me to the MacGyver part of this tale.

Brett was sharing a room at the rather swanky Le Meridien Hotel on Canal Street with me and his best man, Stew Creed. By the time Stew and I managed to drag Brett out of the bar, our Cajun friend couldn't stand up without assistance. Which is why Stew and I had propped him against the double-hung French doors of the Meridien's bridal suite while we attempted to unlock the door to our own room, located directly adjacent to the bridal suite. (Elizabeth, who was spending the night with several girlfriends in another hotel, hadn't yet moved into the bridal boudoir.) As we fiddled with the key, Mouton somehow revived himself, seized the ornate brass handles to the bridal-suite doors, and pulled as hard as he could while yelling at the top of his lungs, "This is where we're stayin' tonight, boys!"

Now, Brett isn't exactly built like a linebacker. But the force of his drunken pull popped the hinges, locks, and plates completely off the latches. And the doors swung open, revealing the interior of the suite in all its splendor.

Since none of us had ever actually seen a bridal suite before, the first order of business, obviously, was to venture inside and check the place out. We whistled softly over the contents of the bar. We merrily clomped up and down the staircase to the canopied, king-size bed. And we conducted a close inspection of the gold-plated fixtures inside the walk-in shower—which was, I think, about the time that the suite's ivory-engraved telephone started ringing.

By this point, Brett had passed out on top of the bed, so I was the one who picked up the receiver. It was a guy down at the front desk, who had somehow learned of our break-in and was calling to announce that he had dispatched a security team to deal with "the situation." I hung up and informed Stew that we had about two minutes to fix this mess, and that if we failed, Brett might be waking up to rays of striped sunshine inside a New Orleans jail cell. On his wedding day.

At this point, Stew remembered the groomsmen's gifts in our pockets.

I'll limit the description of what happened next to simply saying that Stew and I will forever be indebted to whomever designed the Leatherman all-in-one, whose needle-nose pliers, awl, diamond file, and triple assemblage of screwdrivers all played crucial roles in our emergency repair procedure. After ejecting Brett into the hallway, we managed to scoop the screws off the carpet, re-anchor the plates, reassemble the entire latching mechanism on the door, close and relock the doors, and drag Brett into our room next door before the security team even made it off the elevator.

I don't think an Indy pit crew could have done a more professional job—a fact which Mouton, to his credit, has never forgotten.

Each year, just as his anniversary is approaching, Brett picks up the phone and gives me a call.

"Thanks for being there, my friend," he says, muffling his

voice so his wife won't overhear. "I owe my entire marriage to you and Stew."

Yep. Us and a pair of Leathermen.

Kevin Fedarko is a writer living by himself in Santa Fe, New Mexico. He owns neither a dog nor a cat, but is in the process of acquiring a set of very cool power tools, with which he plans to build his own house. He recently committed to a domestic partnership with a potted geranium.

The Underwear

by Renee Dale

THE UNDERWEAR WAS BEAUTIFUL. THIS WAS BEFORE MY PREFER-ENCE for the classic white cotton bikini and so I'd made a collection of silken pairs. These had a black background and a paisley design the color of green jewels. They were not expensive (I was in high school at the time), but they seemed terribly sophisticated to me. I wore them only occasionally, under a perfect-fitting pair of jeans or a whirly sundress. That night I had them on because I was going to a party—an important one with the potential to propel one further into the social stratosphere.

The party was in the meandering Victorian where my friend Melinda lived, a house so large it seemed to tilt against the sky. Everyone I knew or wanted to know was there—a brutally edited group, but perfect for the narrow purposes of a high-school party. There were people on the porch in groups of three or four, people dotting every corner of the blue-green lawn, people under the trees and in the dining room, people sitting on her parents' monastic, prim-looking bed. They were everywhere, spilling drinks, touching one another, laughing wildly while Melinda's parents were off vacationing somewhere; their daughter hosting a just barely controlled evening.

I walked outside to search for my best friend, Jamie, and eventually found her on the lawn. Some easy hours passed as we sat outside, swatting mosquitoes, drinking cups of beer, flirting with friends and strangers.

After a while, I wandered back in and took the narrow kitchen stairway up to Melinda's bedroom to look at myself in the mirror. It felt necessary to make sure I looked passable enough to be having as much fun as I was. When I turned to leave, he was there, sitting on the window seat. "I brought you a beer," he said, holding a brown bottle toward me. Not even a keg beer, but one he had opened himself (his forearm flexing against the opener?!). This was one of *the* boys, a few years older, probably nineteen to my sixteen. He was one of the Legends, as guys my own age mockingly but enviously referred to them.

Even in that old T-shirt his body showed through. He always moved as if in a state of extreme relaxation, his walk so loose it seemed vaguely uncontrolled. I noticed, even beneath the worn khaki pants, that this was incongruous with the forever-flexed appearance of his legs, practically equine in their tautness. He had white, even teeth and a small, deep scar on his cheek.

He led the way through some verbal sparring, which after a

time led to him touching me, which eventually led to him talking about finding some "privacy" so we could "be alone." This
was suddenly urgent. People continued to peek their heads in
and mill around in the hallway, so he stood up and led me, ab-
surdly, to Melinda's bedroom closet. As comic as it seemed at
first, by the time he had me across the threshold, we were both
smiling and leaning into each other's bodies, colluding. It was
better than if he'd ushered me into a lavish ballroom, his hand
at the small of my back. Once inside, he maneuvered us with
real skill (had he entertained in this particular closet before?).
We lay on the floor, the blessedly shoeless floor (Melinda must
have had one of those nifty shoe bags on the door—I never
thanked her for that) with shirtsleeves dangling above our
heads.

For a while—maybe it was two minutes or it could have
been ten—the boy took his time, easing me out of jeans. Then,
in one efficient tug, he ripped my underwear. Tore it right off
my body. Not in a fumbling, groping manner but in a hungrily
efficient, eyes-on-the-prize move that shocked me silent. I was
amazed by this flourish! The usual period of negotiation had
already been hastened by those magical hands, those pink lips,
and now this—the savagely dismantled underwear. It was an
exponential turn-on. Still, I felt a pang of tenderness for myself. Only hours before, I had lifted them reverentially from
their perfect triangle fold in my dresser, blind to their future.

I never wanted to leave the closet. In the light of the room it
would be imperative to conceal how knocked out I was. More
importantly, I'd have to craftily repair the underwear if I
hoped to rejoin the party. And I *had* to rejoin the party, if only
to gloat among those not blessed enough to have been spirited

away to the delightful closet in the corner bedroom. Of course, tossing the underwear was out of the question. They were a souvenir. Plus, what if someone discovered them in the waste-basket—they could be traced back! They had to be rebuilt.

The boy pulled his T-shirt on over his head in that bunched up, mysterious fashion that in my experience was employed by especially attractive males. He looked as competent doing that as he had taking it off earlier, one-handed with a tug between his shoulder blades. We parted with an exchange of pledges and promises, and then he was gone.

Immediately my mind turned to chasing down a solution for mending the torn bikini. The search was confined to Melinda's room. Rifling through junk drawers in the kitchen was impossible. The glances it would have prompted and the discomfort of glissading about down there, trying to look nonchalant in semiattached underwear seemed a crazy tableau. My options were stark. There was no desk in the room which might have yielded considerable booty—any form of affixing object—a rubber band, a lone paper clip nestled there in a lit-tle dugout for the taking.

On top of the dresser there was a sprinkling of earrings, twisted hoops and bent studs, some without a mate. There was a remote control and a blue pen cap with forensic-worthy dental imprints. And then I saw them. A pile of index cards, blank and poised. Why did she have these? For what—flash cards? The prospect was quaint, but made all the more poignant by the fact that the cards were fastened into a stack by a black and silver binder clip. I harvested the thing and made my way to the hall bathroom, moving past people like a ninja.

My legs trembled, not unpleasantly, as I stood before the

mirror. The two strings of fabric hung lamely, sadly even, like a demolished chastity belt on my thigh. I pulled the crimped fabric taut and placed the strings in the jaw of the clip. It dug into my hip a bit, but this was a minor penance to endure for the elation of the closet encounter.

I was downstairs grinning like a fool when I felt the clip lose hold of the underwear with a dull *pop*. I raced back to the bathroom, accompanied by an odd limp, the clip wedged painfully between my jeans and my thigh. I muttered something violent, thoroughly cursing the clip for being a medium size when I'd needed the mini.

This time, the bathroom door wouldn't lock. The mechanism spun like a saucer in the jamb and the burden of this produced a constellation of pink panic splotches on my neck. How could my bliss be forfeited to this predicament? I could have strong-armed the knob into place but I was phobic about getting trapped in small rooms. I set the ball of my foot against the bottom of the door and pulled down my pants. Again. Someone knocked. I reached for the faucet and smacked it on, the innocent sound of running water meant to convey that nothing gruesome or humiliating was going on in there, just a quick soaping of the hands. After a series of frustrating attempts, wherein the fabric repeatedly slipped from the vise, the clip went briefly airborne and landed on the octagonal tiles. It ultimately came to rest in the shadows behind the toilet, a terrain too inhospitable to plum. I capitulated and made the only available correction—the simplest approach, and the one I should have thought of from the start (and would have, had my head not been spinning). I double-tied a neat, tiny bow of black silk at my left hip. Lopsided, but festive.

Back downstairs, I discovered that one of the boy's associates had similarly introduced himself to Jamie. He had backed her into the pantry or the stair landing, some servants'-quarterish space, seedy or sexy depending on your mood. They had pursued us in a calculated way, unaware that we even knew each other. We loved this. Her Legend was the kinder of the two, certainly more gentle. Mine was better looking, with a growing reputation for being pathologically generous in bed.

Almost everyone had gone by the time I returned from rigging my undergarments, so Jamie and I decided to leave, too. She offered me a ride, but I decided to walk. It was humid that night. The air vibrated with busy insects. My arms were slick with sweat and my lips salty. I walked through deserted neighborhoods, cutting across wet lawns and over garden fences, the blue light of television sets spilling out of dark houses into the street.

More than his hands on me that night, or all the times we met after that, I remember one thing with total clarity: I took the winding, macadam path through the woods that usually petrified me, the darkness as solid as a wall, and crossed the elfin footbridge totally unafraid. The proud, perfect bow I had made rested on my hip and rubbed against my jeans as I walked. It felt erotic and triumphant. Legendary.

Renee Dale is a writer living in New York. She's working on a novel and hopes her parents never come across this story.

TROUBLE WITH THE AUTHORITIES

WHEN I WAS A SENIOR IN HIGH SCHOOL, I DESPERATELY WANTED A fake ID. This was 1989 in Annandale, Virginia, just outside the Beltway, and there wasn't much to do on Friday nights except drink beer, blast *Appetite for Destruction*, and attempt to dance like Axl Rose. My friends and I were going to get our alcohol one way or another, and the practice of shoulder tapping—loitering outside 7-Eleven and offering a free six-pack to incoming customers willing to buy us a case of Milwaukee's Best Light—was getting old. Not to mention risky. And, frankly, pathetic. Ultimately, this was a matter of pride. At seventeen, were we not man enough to buy our own damn beer? Were we not resourceful enough to find a way?

It was my friend Phil who came up with a simple scheme that now reminds me of Willie Sutton's famous answer when somebody asked him why he robs banks. ("Because that's where the money is," Willie said.) We would drive to the Arlington branch of the Department of Motor Vehicles, back his gray Honda hatchback up to the Dumpster in the parking lot, and load his trunk with DMV garbage. "That's where the IDs are," Phil reasoned.

And he was right. After driving back to Phil's house and sifting through three or four bags of soggy, rancid trash, we managed to harvest the raw materials for twenty Virginia driver's licenses. The main things we needed were the perforated green cards on which each driver's info was printed. But we also needed the clear plastic sleeves that held the cards. (This was before Virginia started laminating its licenses, making things far more difficult for underage drinkers all over the state.) We found plenty of sleeves but only twenty green blanks that had managed to stay dry in the garbage. Still, Phil deemed the operation a success.

We opportunistically recruited a friendless computer geek named Hal to match the font and format the information on his Commodore 64, then print the cards and mount the photos just so. (He was paid with an ID of his own and some quality time with the semicool kids.)

Phil was the first to try his ID in the field (we had to change 7-Elevens so as not to be recognized)—and it worked. But Phil was a big guy, six-foot-two and well over two hundred pounds, with a passably mature-looking mug. Would it work for me, with my slighter build and baby face? Yes. For Nate, whose blond curls made him look like an overgrown cherub? Yes. For Walter, who stood five-four on a good day? Yes! We were in business.

Literally—and that was our downfall. We decided to sell the fifteen remaining IDs to our classmates. Hal was again recruited to do the grunt work. (We did cut him in on the action.) We charged fifty bucks apiece and sold out within a week. That's a lot of beer money, especially when your brand costs $6.99 a case. We were golden—until a few weeks later, when Mr. McGowan, a retired Army man who now taught shop and doubled as an administrator, overheard one of our customers bragging about his new ID. I never did hear a convincing account of what happened next, but suffice it to say the kid did *not* MacGyver his way out of his predicament. In fact, he sang like the Mormon Tabernacle Choir.

We were finished. It seemed tragic at the time, but now I realize the school went easy on us. All we had to do was return our IDs and rat out all fifteen of our customers, and the charges would be dropped. It shames me to admit it, but we took the deal. (We may have sang, but we weren't exactly Sopranos.) And then we went back to shoulder tapping.

Bummer, huh? Well, yeah, I guess. When I was seventeen I certainly felt that way. But now I'm glad it happened. I'd rather have the story now than the beer then.

And that's the thing about unexpected run-ins with the powers that be. Below a certain threshold of seriousness—say, jail time and/or a fine of five hundred dollars or more—they almost always yield stories that justify the hassle.

Like these.

This One Time, at Band Camp

by D.W. Martin

THIS IS THE STORY OF TWO SAVVY THIRTEEN-YEAR-OLDS AND HOW they almost masterminded the cover-up of the century. Or at the very least, the cover-up of spring 1991.

As a freshman in high school, I was quite dorky. I was short, fat, and wore glasses. My best friend, Leonard, was shorter, squiggly thin, and wore braces. I don't mean to shock you, but we were in the band. I played the French horn and Leonard played the clarinet. Pretty sweet.

As dorks, Leonard and I were responsible, reliable, and easily prone to adult intimidation. Teachers trusted us. We simply didn't have the courage to do anything bad. And that's why, as mere freshman, on our spring band trip to Myrtle Beach, our band director bestowed upon us an awesome responsibility: She appointed us the Key Masters of our suite in the hotel—one of those garish pink stucco jobs that attract short-tempered fathers, their screaming children, and high-school bands. Each suite housed six guys or girls, two of whom were in charge of the room keys. The Key Masters were also responsible for making sure the suite didn't get trashed.

This really shouldn't have been too difficult, but there was a malcontent in our suite. His name was Bill, he was a junior, and he was not happy that two freshmen were the Key Masters. Bill played in the percussion section, which surprised no one seeing as how he loved to bang things that were not drum-related (such as people's heads). When we got off the bus in Myrtle Beach, he ordered us to give him a key or he'd "open a can of whup-ass" on us.

Bill was the lost member of the Manson family. He was a cross between a rabid ferret and a sociopath. His eyes were pitch black, the same color as his teeth. He didn't so much speak as spit. He wore camouflage mesh baseball caps and hunting T-shirts and it seemed as if he had a lump of chewing tobacco, or "chaw," as he put it, surgically attached to his jaw. Leonard and I were like two floppy-eared bunnies in the den of a wolverine.

Obviously we planned to stay far away from Bill and allow him to do whatever he wanted so long as it did not involve violating us. This worked for about the first thirty minutes, until we heard the door to our suite bust open, followed by an "Oh, crap." It turned out that Bill and our suitemate Noah (a spastic freshman) had been involved in a Super-Soaker skirmish. (Remember Super-Soakers? They were oversize water guns, and really cool if you were a nerdy, undersexed teenage boy.) Anyway, at some point during the fighting, Noah locked Bill out of the suite by sliding the metal safety latch at the top of the door into the locked position. Bill did the only sensible thing: He kicked the door in. Fog of war, you know?

Somehow, though, the door seemed to be okay. It still opened and shut, but the latch and the wood paneling on the door-jamb (where the latch locked into place) had been knocked off. It was a distressing sight. The pieces, which were still latched together, lay on the floor like an amputated limb, and the door-jamb now sported a hideous bald spot of nonlacquered wood where the paneling had been. The only damage to the door proper was a missing sliver of wood where the latch had been glued on (instead of, say, screwed on—safety first at this Myrtle Beach hotel).

The first thing Bill said after viewing the scene was "If anyone tells anybody what happened here, I will kill them. And their families. Is this understood?"

Leonard and I were reluctant to quibble with Bill on this point, but we did think that, perhaps, in this very unusual case, it would be best to tell the truth. *How do you overlook a busted door?* we argued. Plus, we were the Key Masters. The room was our responsibility. And we weren't about to let the adults down. Bill's rebuttal, however, involved flecks of tobacco raining down on my face and a suggestion that my body might be better served with two assholes. Tough to argue with that.

The four of us stood in the hall of the suite looking at the debris on the ground. How, exactly, were we going to cover this up? Bill thought we could try to glue the pieces back into place. Or maybe we could say it was like that when we got there, and that we somehow overlooked it when we first inspected the room. Another idea was to throw away the broken bits and paint the doorjamb a matching color. (Not that we had access to glue or paint or any tools whatsoever, but Bill liked to think big.) It was becoming apparent that he needed our help. And if we were going to go to the trouble of covering this up, what was the point of embarrassing ourselves? Time for the two smartest guys in the room to devise a foolproof plan.

Our solution was quite simple. We would fit the broken pieces back into their original places, like a jigsaw puzzle, then lock the latch. This actually worked well. When we pounded the pieces back into place, we were surprised by how firmly they took hold. Not quite good as new, but close. We knew, of course, that the fix wouldn't hold up under scrutiny, and our plan accounted for this. Later, shortly before the chaperones began making their rounds, we would secure the latch, but position the door so that it was ever-so-slightly cracked open. Then, when the chaperone knocked, one of us would casually say, "Come in! It's open!" And when he opened the door, "breaking" the latch, the broken pieces would come off (again). We would all have a good chuckle, shake our heads at the South's shoddy carpentry skills, and get back to free HBO.

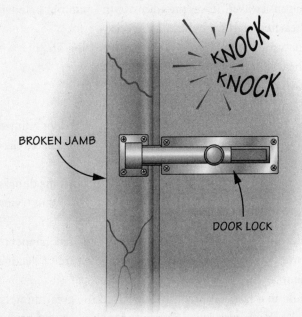

We rehearsed the plan, tested it out, and deemed it flawless. Who knew Leonard and I could be this devious? Who knew how thin the line was between dorky and diabolical? I even thought I detected a glimmer of respect, or at least an absence of murderous rage, in Bill's eye.

Soon it was 8:30 P.M. Chaperone time. Go time. Leonard and I sat on the couch, watching Nick at Nite. Bill, with all the care of an artisan putting the finishing touches on a Fabergé egg, placed the hardware and wood slivers back into place and gently slid the latch into the locked position. Then he ran into the TV room where we all sat, breathless. Our hearts pounded in unison. The adrenaline flowed. So this was living! This was a life of intrigue and danger!

"You guys awake?" It was the voice of Mr. Fouts, our balding, good-hearted, slightly chubby chaperone.

"Yup, come on in," I said. "We're just watching TV."

We heard the door moan.

"Seems to be locked," said Mr. Fouts.

"No, must be stuck. Just give it a little push."

Those pieces held their ground. He grunted. The door grunted back. And then the crash of wood and metal as the pieces clattered across the floor.

"Oh no," Mr. Fouts said. "I seem to have broken the door."

We leapt off the couch and into the hallway, each of us trying his best to act stunned.

"What did you do Mr. Fouts?" "How did that happen?" "Jeez, you've got quite a grip there. I better be careful shaking your hand."

I made that last comment. Honest. And the great thing is, he bought the flattery. He really reveled in the idea that maybe hidden inside his middle-aged potbelly was a Hulk Hogan just bustin' to come out. I remember him saying, "I guess I don't know my own strength. Wow, I just knocked down the door."

We told him that we had latched the door during the Super-Soaker battle and had forgotten that it was still fastened. He seemed a little skeptical—I need to give him that—but he was also more than happy to just chalk this one up to "another crazy thing that happens in Myrtle Beach." We were in the clear.

Until Mrs. Morrison showed up. She was the other chaperone. And the instant she saw the door carnage, she knew something was amiss. We tried to explain that Mr. Fouts was just *really* strong. Mr. Fouts nodded in agreement. He *was* super strong. But Mrs. Morrison was a hardass and didn't take shit from anyone. (When her daughter graduated from high school, the poor girl received an award from the school superintendent for never having missed a day of school from first to twelfth grade.) Mrs. Morrison informed us that she would "get to the bottom of this by tomorrow." Which she did. Under intense interrogation, Bill folded. Perhaps he wasn't that tough after all. He ended up paying eighty-five dollars to repair the door (a huge sum when you're a teenager, the rough equivalent of $750 in adult money) and got two weeks of detention.

Beautifully enough, though, no one believed that Leonard and I—the good kids—could have been behind the plot to fool the chaperones. We paid five dollars in damages and apologized. And learned to never use our immense powers for evil again.

D.W. Martin is the author of Officespeak, *a comedic look at, you guessed it, the language of the office. He lives in Queens, the most magical place on earth.*

The Crossing

by Jason Sheftell

WE WERE DRIVING FROM AMSTERDAM BACK TO PARIS, OUR HOME for the year, racing to return the rental car within the twenty-four-hour limit. Four silly young Americans fresh out of college, sandwiching eight hours of pot-fueled debauchery between sixteen hours of driving. And we planned to bring some of that debauchery home with us. Before leaving, we had opened the windshield-wiper-fluid container and dropped in ten little baggies of Thai Stick wrapped in tiny tinfoil packets. Smart, we thought. Well hidden. We weren't dealers, just kids tired of paying retail in Paris.

We reached the Belgium-France border at about five in the morning. It was still dark, and the two groggy fools in the backseat couldn't locate their passports quick enough for the guards in the tollbooth. They told us to pull over to the side, then escorted all four of us into the station house.

In a small, barren room decorated only with a few posters of French tourist destinations, a six-foot-five border gendarme—that's French for "scary cop who held a cigarette in one hand and my fate in the other"—snickered, grunted, and began his interrogation.

"Where were you?"

I was the only one who spoke French. I had to answer. "Brussels."

"Why?"

"Research for school."

I was quivering. This was the end. My parents were gonna flip. And that's *if* I was allowed to contact them and ask for help. Maybe I wouldn't even get that chance. Maybe I'd just rot in a French jail, surrounded by prisoners calling me "*mon chéri.*"

"One day?"

"Yes. I have a paper due tomorrow." A lie. I wasn't even in school.

"Empty your pockets!" he screamed. "All of you! Now!" The two guards stepped forward.

We had actually thought of this and emptied our pockets of Dutch currency before we hit the French border. Just in case. Only wait. The slower of the two dolts in the backseat—slow to find his passport, slow in general—looked like he was going to cry. When he reached into his back pocket, a fistful of Dutch guilders spilled out.

"Amsterdam," said the gendarme, grinning with malevolence.

The guards approached my friend and whisked him through a doorway in the back of the room. The rest of us, all now sweating, were taken outside to watch a German shepherd nose through our rented Renault.

A pair of mechanics removed the doors and ripped them apart. Nothing. Then they put them back together, which was very nice, considering. They opened the trunk. Two books on Rembrandt that I'd bought at an art bookstore in Amsterdam. The gendarme looked astonished. He picked one up and started flipping through it.

I began to prattle, almost beg.

"*Je te dis que je suis etudiant,*" I said. I told you I studied. We smoked it all in Holland. I swear. We did. We, I, would never risk my life on something so stupid. Please. There is nothing. Yes, we went to Amsterdam. We got high but brought nothing back.

I'd never spoken such fluent French in my life.

He wasn't buying it. He went to the hood. My friend, the ringleader and driver, looked down and shook his head. The gendarme turned on his flashlight. He pointed it down at the engine and peered. His scaly, pockmarked skin shone in the light. He aimed the beam at the little tank that held the wiper fluid. He unscrewed the cap and bent down to examine it. He turned to one of the mechanics in orange.

"*Liquide, c'est tout.*" Just liquid.

The tinfoil packets had sunk to the bottom. He couldn't see them. And the dog couldn't smell the Thai Stick through the chemicals. We were saved.

Our friend with the guilders emerged from the station house looking pale and mildly traumatized. He walked over to us and found something fascinating to stare at on the ground between his shoes.

"*OÙ EST LE SHIT?*" the gendarme demanded, louder now, the frustration rising in his voice. Where's the shit? I know you have it, he said.

Our ringleader just shrugged. He knew enough French to understand the question, but he pretended like he didn't. I said nothing.

The gendarme stared back at us with a mixture of disgust and boredom. "*Au revoir,*" he finally said, dismissing us with a wave of his hand.

We piled into the Renault and drove away, staying well below the speed limit, at least until we were out of sight.

In the car, we were overjoyed. Giggly. Except our friend with the guilders. He looked sullen.

"What happened?"

"Strip search."

"Like, cavity search?"

"I don't ever want to talk about it."

We were silent the rest of the ride home. Our absentmindedness—not emptying the fluid before inserting the packets—had saved our young lives, or at least saved us a trial for drug smuggling: When we reached Paris, we removed the packets before returning the Renault. The weed was too wet to smoke, of course, despite our lame attempts to seal it. We let it dry for two weeks, then we smoked it anyway. We never had such headaches. And we—or at least I—never tried anything that stupid again.

Jason Sheftell is a New York writer who committed one or two other very minor infractions (like jumping the subway turnstile) while living in Paris in the late 1980s. Though he's almost forty now, his parents are still going to flip when they read this.

The Vodka, the Blood, and the Miraculous Coffee

by Evan Rothman

I WAS DRUNK—VERY DRUNK. SO DRUNK, IN FACT, THAT WHEN MY friends dragged me out of the Greenwich Village bar for trying to pick a fight with a six-and-a-half-foot bouncer who had looked at me funny for stumbling into him, I was holding two double vodka shots, both for myself. I was also, of course, recently dumped and burrowing further into the dumps. Post-collegiate blues, 1992.

Once outside, I continued to rant and rave about this cretin who'd dared look at me sideways for an honest mistake in equilibrium, bringing myself to a righteous froth, when suddenly I was no longer upright. In fact, I was facedown on the sidewalk. The culprit? A fire hydrant in my (wobbly) path.

This sent me into a fit of laughter; my friends were amused, too, until they saw the blood coursing down from the gash where my forehead had conked the concrete. Then a well-dressed gay couple appeared on the scene.

"He needs coffee," said one to the other, who nodded and then set off, presumably to try to find a diner still open at 1 A.M.

The man returned a few minutes later, not with an extra-large-black-no-sugar but rather with coffee grounds, which he rubbed into the cut with brisk, no-nonsense efficiency. Presto—the bleeding stopped. Coffee, as it turns out, works as a coagulant. My friends, dumbstruck, thanked these Good

Samaritans as they went on their way, presumably to rescue other imbeciles in need.

I would love to say that the coffee grounds also sobered me up. Alas, that process took several more hours. One thing I do know: They must have used decaf, because I slept like a baby that night.

Evan Rothman, a freelance writer in Brooklyn, has always been a lightweight when it comes to drinking. The former executive editor of Golf Magazine, he now confines his alcohol consumption mostly to the golf course.

Seventh-Period Stretch
by Joshua Foer

NINETEEN NINETY-SIX, THE YEAR MY BELOVED BALTIMORE ORIOLES finally shed their hapless loserdom, coincided with my freshman year of high school, when I rushed headlong into mine. After more than a decade of dwelling near the cellar of the American League East, the team had qualified for the playoffs as a wild card and advanced to the American League Championship Series. Their opponent? The despicable New York Yankees. I wore braces and Ray-Ban Wayfarers.

In the Foer home, where a large framed Cal Ripken poster hung in the basement opposite the wall of family photos, this was a moment of considerable excitement. (Well, for the men of the family it was—my mom was "happy we were happy.") All those years of unrewarded devotion, all those Yom Kippur benedictions, had finally amounted to something. I understood that the rest of my life after that October would simply be denouement.

The problem that threatened to polarize my family? Several games of that Yankees series were day games, played during school hours. My father, who could count the number of O's games he had missed that year on three hands, was prepared to let me stay home "sick" so I wouldn't have to suffer the indignity of watching the games on tape delay. My mother, who'd used her box seat at the stadium as a reading chair until my dad had stopped inviting her to games, thought this would set some sort of bad precedent. Or something.

"No means no," I remember her telling me, though I suspect it was Dad she was really addressing.

Family rap sessions were held. Tears may have been shed. Desperate measures were considered. I called the local counterespionage store—I grew up in Washington, not Baltimore—and explained my predicament. I asked if they sold a concealable radio that would allow me to attend school and listen to the games at the same time. Maybe something with one of those Secret Service earpieces? The answer was no, at least within the budget of my allowance.

In the end, Mother was the necessity of invention. She was holding fast against any scheme that involved me missing school. So, a day before the series started, while she was out painting ceramic mugs at a Made by You store, my father and I hatched a plan. We tore apart a pair of the smallest headphones we could find and superglued the innards of one earpiece to the underside of my oversize wristwatch—which, despite my dropped voice, still had mathematical functions. A sweater far too warm for D.C. in October concealed the cord that snaked up my arm, under my armpit, and down to the FM receiver in my pants pocket.

I sat through world history with my hands clasped behind my head and the watch radio whispering play-by-play into my ear. Before class, someone had written the score on the chalkboard: O's 2, Yanks 2. That changed in the top of the third when Rafael Palmeiro smacked another one of the towering home runs he'd been hitting all year. While my teacher lectured, I flashed the new score to a friend in the front of the room. He got up and corrected the chalkboard. My cell phone (yes, I was an early adopter) was already buzzing in my pocket. It was my dad, calling to make sure I'd heard.

Joshua Foer was raised on a diet of Mr. Wizard, Quantum Leap, and MacGyver reruns. He is now a freelance writer in Washington, D.C.

Confessions of a Preteen Arsonist
by Donald K. McIvor

I HAVEN'T BEEN ABLE TO UNEVENTFULLY MAKE A HOTEL RESERVATION since 1985. That's when *MacGyver* premiered, and guess how you pronounce my last name? I always get a chuckle when, after the clerk makes the connection, I relate how I can fashion an atomic weapon out of paper clips or whatever.

Back in the 1930s, however, when I was growing up in East Kildonan, a tiny suburb on the outskirts of Winnipeg, Manitoba, not only was there no *MacGyver*—there was no TV. Not to mention no computers, video games, or Xbox 360s. On the Canadian prairies, life was pretty simple. So kids like me made our own entertainment and adventure. This story relates how a buddy and I amused ourselves with a tricky little contraption that, now that I think about it, probably would *not* have made our hero proud (and which I don't recommend you try at home).

Here's how we made it:

1. We got two standard wooden spring-loaded clothespins.

2. We took one apart.

3. Then we forced one of the wooden halves of the disassembled clothespin through the spring of the other.

4. And finally, we used the loose spring to clip the whole thing together.

We could now cock the spring like the hammer of a rifle. The bullet, in our case, was a standard strike-anywhere wooden match that was wedged into the jaws of the contraption, with the match head facing the shooter. With a flick of a thumbnail, the cocked spring would snap forward to light the match and send the flaming projectile twenty-five or thirty feet.

One mild September day, my friend Sam and I were riding our bikes along a gravel road, firing lit matches into the grassy ditch to our right. No particular reason, of course, just a couple of bored boys engaging in some random pyromania. Usually this would burn the grass out of the ditch, but not much more. This time, however, the wind carried the fire up out of the ditch and into the tinder-dry grass of the adjoining field. We got out of there in a hurry and hid to watch the fire eat through the field.

And that's when we noticed that a wooden frame for pouring the foundation of a house had been built in the middle of the field, directly in the path of our blaze. We hadn't seen it at first because it was hidden by the high grass. Too late now. As the frame was consumed by the flames, we knew we were doomed to life behind bars. Not that we planned to turn ourselves in. Instead we slunk back to our homes by way of back routes and destroyed all evidence. I was petrified, of course. Not only of law enforcement, but of how my parents would react—would the good name of the clan McIvor be forever sullied by the criminal acts of one of its junior members?

Each time we saw Chief Coley, the lone policeman in our town, we behaved furtively. But as the weeks turned to months and then to years, we realized we'd gotten away with (accidental) arson. We never did find out whose would-be house that was. But ever since, I have been very careful with matches.

Donald McIvor was able to rise above his early life of crime to become a geologist and oil-company executive. Now retired, he is the author of Curiosity's Destinations: Tales and Insights from the Life of a Geologist *(Grindstone Press, 2005).*

Pass the Potato

by Donnie Snow

IN THE LATE '90S, BEFORE THE DOT-COM CRASH, AN INTOXICATING mood of Clintonian prosperity allowed otherwise-modest young liberals to indulge in the type of lavish, guilt-free revelry more common to the supply-side P. J. O'Rourkes of the world.

As a grad student in Baltimore at the time, I'd found a small collection of these people for friends. We all believed in quality debauchery, but had reached the age when waking up on somebody else's floor and breakfasting on warm beer and cold pizza was, frankly, unacceptable.

These generally higher standards for comfort and quality led to an upgrade of our traditional New Year's Eve bash: The liquor still came in a bottle with a handle, but the beer was microbrew, the food was prepared instead of purchased, the cigars didn't have plastic tips, and the ritual sacrifice of empty kegs was altogether discontinued.

Even the marijuana was high-end. Bobby, a future attorney, had just scored a bag of grade-A hydroponic from a Grenadian short-order cook. But somehow we'd all become so adult that nobody could scramble up so much as a pipe or even rolling papers. And I'd long ago jettisoned the port-a-bong stowed in my VW microbus.

Crowded into a narrow apartment kitchen with Bobby, a Czech expat, and Frank, an environmental lobbyist who was hosting the party, we exchanged incredulous gazes, then tried to decide what would be worse: rolling it up in newsprint or asking strangers at the party if they had anything we could use.

Before Frank could get to his recycling bin and yesterday's classifieds, I remembered the potatoes I had brought to make my famous hangover huevos for brunch tomorrow.

"I'm all over this," I said, reaching into the pantry for a spud.

Bobby handed me his Montblanc and offered some advice: "No need to be an artist. It's all about airflow."

How to Fashion a Potato Bong with a Pen and Aluminum Foil

1. With the pen, core a potato in two places—one hole on top, another from the side—until the two shafts meet.

2. Using the tip of the pen, poke a tight array of tiny holes into a small swatch of aluminum foil.

3. Tuck that foil into the top shaft.

4. Drop in a bud, fire that mother, and inhale through the side shaft.

5. Pass the hot potato.

Long past midnight, after the whiskey had run out and the potato smelled like the Hudson River, Bobby cornered the three of us. "I have come to the realization that the convictions of my youth are on a collision course with my preferred lifestyle," he lamented. "On one hand, I believe in equal pay for equal work,

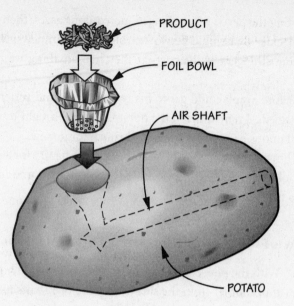

PRODUCT

FOIL BOWL

AIR SHAFT

POTATO

a coherent long-term environmental policy, and safe highways. But then again, I don't want those things to interfere with making a shitload of money, reaping the benefits of exploited child labor, and driving a fast car down the interstate with some hottie who may or may not be my wife."

We grunted in empathy.

This went on for awhile, each of us chiming in, until—drunk, belligerent, and increasingly resentful about approaching thirty—we did what most guys in our position would do: We broke into one of the Johns Hopkins frat houses across the street.

**Three Things to Remember When Raiding a Frat House
to Recapture Your Youth**

1. Bring beer. If stopped by a cop on the fraternity's property, you can say you're late for the party; if caught by a
   frat brother, well, you've brought gifts.

2. Use the front door. Frat houses are rarely locked. You can
   usually walk right in like you own the place.

3. Confiscate any beer from the fridge and at least one useless, miscellaneous item for sport.

We heard some rustling upstairs as we liberated thirty-two
cans of Schaefer and Milwaukee's Best, along with the frat's
framed yearbook photo from 1998, but the house seemed
mostly empty. It was the middle of winter break, after all, and
most of the brothers were home for the holidays.

There's no telling exactly how much time passed before the
lingering frat guys woke up and noticed their missing 3' x 4'
photo or crappy beer, but the cop who answered their call took
only a few minutes to notice the Bacchanal raging across the
street.

It must have been past 3 A.M. when the officer knocked on
the door. Frank invited him in and poured him a beer (hey, it's
Baltimore) while everyone else tried not to look at the pool
table where we'd hidden the photo under the cover.

By the time the cop finished his cursory inspection of the
apartment, I was slicing the potato into the sausage gravy.

"Hungry, Officer, uh, Levy, is it?" I asked, proffering a biscuit.

He took the biscuit and drank the beer, but passed on the gravy.

We all did that year.

Donnie Snow is a freelance reporter who lives in Memphis. He still brags about his famous huevos.

ON THE JOB

WHEN I GOT MY FIRST BIG PROMOTION BACK IN 1997, MY DAD wrote me a letter. It was a beautiful letter, full of love and pride, and it was designed to both congratulate me on what I'd accomplished and prepare me for what was to come. "Never compromise your integrity," he wrote. "No matter how dire the situation, keep your cool." "Practice empathy." Yes, it was the advice of millions of fathers to millions of sons—and millions of senior executives to millions of middle managers—but this was *his* advice, *his* hard-won wisdom, and he'd taken the time to boil it down for me, one nugget at a time. I still pull that letter out and read it every once in awhile, and in twenty years or so, when, God willing, my own son finds himself in a position of responsibility, I intend to steal from it liberally.

But there was one line that I remember more vividly than the others, and that's because I found it so surprising: "Act confident, even if you're not. No one can tell the difference."

Really?

Dan Vaughan, a banker who stayed in his first job for more than twenty years, an Irish Catholic who went to church every Sunday, a company man who wouldn't call in sick unless he was literally unable to move, was telling me to . . . mislead?

Of course not. Now, almost ten years later, I understand what he was really saying: Sometimes you gotta be MacGyver on the job.

Not in the traditional sense. My dad has no gift for the mechanical arts, and because he was never able to teach me how to work with my hands, neither of us expected me to wind up working with my hands. What he meant is that in almost any job, the single most important qualification is the ability to just figure it out, even—especially—when you have no idea how. You need to be cool, no matter how hot the situation. You need to be (or at least look) unflappable. Confidence, after all, is the first step toward solving any problem.

In one way or another, all of the characters in these stories followed my dad's advice. Sportswriter Chris Jones had no idea how (or even whether) he would find Ricky Williams, the Miami Dolphins running back who'd fled to Australia, but that didn't stop him from assuring his bosses that he'd pull it off. A.J. Jacobs, who wrote a book about reading the entire encyclopedia, came up with a brilliant, preemptive scheme for dealing with the hecklers he knew would hound him on his book tour. And Stacey Grenrock Woods, well, Stacey's workplace MacGyverism may be the most genuinely clever (and entertaining) story in this entire book.

We begin at the beginning, with Susan Casey's story about MacGyvering herself into her very first job. I doubt my dad has ever done anything so daring, but he'd definitely applaud her chutzpah.

MacKirk

by Susan Casey

As anyone over the age of ten can tell you, the world is a competitive place—and sometimes extreme behavior is required in order to get what you want. Nothing as radical as, say, annexing a country or knocking off a bank, just a subtle twisting of the rules from time to time in order to achieve your goals. One such skill that I have found especially useful over the years is the ability to angle my way into almost any job.

I began to develop this technique largely out of desperation, just a few weeks after I graduated from university. I was perusing the classifieds in the *Toronto Star* when I spotted an ad for my dream job: an entry-level position at a hip Toronto advertising agency, a career-making place where even the briefest employment would open future doors. I had no intention of letting anyone else get this job. Pulling together my first résumé, I carefully matched my "career objectives" to the job requirements spelled out in the ad. Their "must work well on deadline" translated to my love of "fast-paced challenges." I was working my way down the list of attributes when I hit the big snag: "extensive experience operating a stat camera."

This was the pre-desktop-publishing era, back when mock-ups were created with X-Acto knives, cardboard, wax, and, occasionally, drops of blood. Cumbersome photographic equipment was deployed to produce stats, which were images—either a chunk of type or a piece of artwork—that were printed on photographic paper and pasted onto a board to make a proof for the printer.

The stat camera was a lummox of a machine. Standing at least five feet tall, sporting needles and gauges and glass plates and dials and a vacuum that made a scary sucking noise, it required its own darkroom. The art department at my university had contained such a beast, but the camera was considered too valuable and complicated for experimentation by students, even art students like me. Instead, it was manned by a laid-back, long-haired technician named Kirk. Though Kirk was barely older than any of the students, he'd had an aura of technical competence that set him apart. I called Kirk.

"That's easy," he said. "There are only two types of stat cameras: Agfas and Kodaks. They're *completely* different. So tell them you've had extensive experience working a stat camera, and then once you get the job, if it's an Agfa say 'Oh, I learned on a Kodak,' and if it's a Kodak say 'Oh, I learned on an Agfa.' They won't fire you over that, and it'll explain why you've got no clue how to work the thing."

"Yes, I've spent plenty of time around stat cameras," I said during the interview, which took place in an intimidatingly chic, sun-filled conference room in the agency's Toronto headquarters. "We had one in our department and we used it constantly."

Technically, this wasn't a lie.

"So you'd be comfortable operating one, then?" asked Stefan, an artfully disheveled, thirtysomething senior associate.

"Yes. Absolutely."

This was enough for Stefan. He offered me the job and I took it immediately, giving him no time to change his mind.

On my first day of work, Stefan and I walked down a flight of stairs toward a heavy, ominous-looking door: the darkroom. *Agfa, Kodak, Agfa, Kodak.* The door, which had a thick bolt, swung open. In the auburn light I could make out the silhouette of a hulking machine, taller than my head. This was it: the moment of reckoning, time for me to pull the trick and make it work. I leaned toward the camera. It was an Agfa.

"I learned on an Agfa." The second I said it, I realized my mistake. Unbelievable. "I mean, I . . ."

"Great!" Stefan handed me a large docket of print ads that needed to be created on the stat, all marked with various percentages and instructions for the proper exposures. "Go to it!" And then he swung the door closed and left me in the dark.

And now we enter antiMacGyver territory. As my eyes adjusted to the dim light, I looked around the room. Tubs of chemical baths rested on a table; a rack held drying prints; one wall was dominated by a cabinet that contained a warren of drawers and cubbyholes. Surely there would be an instruction manual in there somewhere? I opened a cupboard—no instruction manuals. But there, sitting on the counter staring at me, was a phone.

I called Kirk.

Speaking in a whisper, I explained my situation. Kirk laughed, sort of nastily. He was used to students botching things up.

"Stand in front of the machine," he said. "Do you see a power switch on the upper left, right beside the image plate?"

I did.

And then he walked me through the entire process. As I operated it, the camera made a grinding and moaning noise not unlike Mack trucks mating. "Totally normal," Kirk assured me. I pulled a stat; it looked reasonably professional.

"Okay, there's one last thing," I said, after forcing him to stay on the line while I shot another dozen. "I've been in here for an hour. Isn't this supposed to take five minutes?"

AGFA STAT
MACHINE

AGFA? KODAK?

"Just tell them you weren't used to this brand of photographic paper," Kirk told me. "And that you had to experiment with aperture readings. And ask them, 'When was the last time you had the lens calibrated?'"

Kirk's advice worked perfectly. Though Stefan did raise an eyebrow at how long it took, he seemed pleased with the images I'd made. And the next time he sent me in to use the Agfa, I was finished in five minutes.

As I've polished my job-acquisition technique over the years—I've accepted and improvised my way through gigs as a bartender, waitress, French teacher, and yoga instructor, among others—I've come to realize that Kirk's lesson alone was worth the price of a university tuition. The trick is not to flat-out lie—no false claims about neurosurgery degrees or helicopter licenses, please—but rather to elegantly bluff, to display confidence in the face of a challenge.

It boils down to a three-word philosophy: Always say yes. After all, in the age of cell phones, you're never far from a lifeline.

Susan Casey is the development editor for Time Inc., and the author of The Devil's Teeth: A True Story of Obsession and Survival Among America's Great White Sharks. She's Canadian, which explains her use of the word "university" in this story.

The Know-Enough

by A. J. Jacobs

HERE'S THE PROBLEM. IN 2004, I WROTE A MEMOIR ABOUT HOW I read the *Encyclopedia Britannica* from A to Z—every one of its 44 million words. I called the book *The Know-It-All*. So, when it came time for my book tour, people kind of expected me to, well, know it all.

It's in the damn title. I mean, Bob Dole wouldn't have written a memoir called *One Soldier's Story* if he had spent World War II milking cows in Kansas. My tour was scheduled to begin in September, so for the entire summer, I studied furiously. I knew that I'd get quizzed at every reading, on every radio interview, at every cocktail party. And I knew I needed to be prepared for whatever someone might throw at me, whether or not it was in the encyclopedia. I bought a heap of almanacs and trivia books to supplement the *Britannica*, and tried to memorize every state capital, every British monarch since Alfred the Great, every species of marsupial.

Problem is, there are a lot of freakin' species of marsupials. The whole thing was an impossible task, akin to drinking and retaining the Atlantic Ocean (from the Greek for "Sea of Atlas"). A lifetime of taunting seemed inevitable. "What's the matter?" they'd say. "Don't know the first bishop of Iceland? You're not a know-it-all. You're a know-nothing!"

Then, one day in late August, about a month before the tour began, I had a flash of insight. It happened when I was at a get-together with my wife Julie's entire family—mom, dad, brother, and various nephews and nieces. We were all in Julie's parents' living room, watching *Romeo + Juliet*, the funky MTV version directed by Baz Luhrmann. During a snack intermission about halfway through, Julie asked me when Shakespeare wrote the original *Romeo and Juliet*, the one without the Mexican mafia gangs.

At once, everyone turned to hear my enlightened response.

"Um . . ." I scoured my brain. Nothing. I scoured some more. Still nothing! I couldn't remember the date of a single Shakespeare play, much less the date of his most famous romantic tragedy. In fact, I couldn't remember much about Shakespeare at all, except that he was born in Stratford-upon-Avon and had longish hair. Oh, and one other fact, which I figured I might as well share. "Well, I do know that Shakespeare and Cervantes died on the exact same day: April 23, 1616."

"That's odd," said my mother-in-law. At which point commenced a lively discussion of death-related date coincidences (Thomas Jefferson died on the Fourth of July, Aldous Huxley died the same day JFK was shot), and everyone forgot that I was totally ignorant of when *Romeo and Juliet* arrived on the scene.

This was a stunningly successful bait and switch. And it led to the most important insight of the whole project: To be a know-it-all, you don't actually have to know it all. You just need to know one remarkable fact about every topic under the sun (a heavenly body, by the way, that is 330,000 times the mass of Earth). I made a mental note to take full advantage of this while on tour.

As it turns out, the trick saved me from humiliation on

multiple occasions. Whatever the question, I always had *something* to say—even if it wasn't 100 percent on-target. During one reading, for instance, I was accosted by a pesky attendee who was intent on stumping me. "Who was the pharaoh after Tutankhamen?" he asked.

"It's funny," I replied. "I was just thinking about the ancient Egyptians. Did you know that they made mummies of their cats—but they also made mummies of mice so the cats would have something to eat in the afterlife? Very considerate."

Sure, people might figure out you're dodging questions like a cabinet nominee testifying before a Senate subcommittee. But if you're interesting enough, they just won't care.

Here's a handy chart of other fascinating bits of trivia, so you can be a conversational MacGyver too:

Topic: Philosophy
Fact: French philosopher René Descartes had a fetish for cross-eyed women.

Topic: Theater
Fact: George Bernard Shaw sat for a very early nude photo in the pose of Rodin's *Thinker*.

Topic: Medicine
Fact: Bayer aspirin invented heroin.

Topic: Film
Fact: Humphrey Bogart coined the phrase "Tennis, anyone?" when he was a Broadway actor playing a rich guy.

Topic: Sex
Fact: Elephant copulation lasts twenty seconds.

Topic: Books
Fact: Edgar Allan Poe married his thirteen-year-old
        first cousin, making him the Jerry Lee Lewis of
        his day.

Topic: Sports
Fact: Touchdowns used to be worth two points, and
        field goals were five points.

Topic: Politics
Fact: George Washington's false teeth were not made of
        wood. They were made of human teeth and ele-
        phant ivory.

A. J. Jacobs is the author of The Know-It-All: One Man's Humble
Quest to Become the Smartest Person in the World. *He does
not know the name of the first bishop of Iceland.*

Bill's Big Day

by Holly Tominack

I WORK AS A REFERENCE LIBRARIAN FOR A CRIMINALLY UNDERFUNDED
public library system in Baltimore City, and anything—
anything at all—that improves morale is very important to our
small staff. One such morale booster is the office birthday party.

Perhaps I should explain what passes for a "party" in my workplace. On the big day, one of us drags out a rickety folding table and covers it with a well-worn, light-blue tablecloth. Then we arrange a cake, some chips and cookies, and the cards and gifts on the table, and someone prints out a Word document that reads HAPPY BIRTHDAY, WHOEVER! and tapes it to the tablecloth. The birthday person cuts the cake around noon or one, and then everybody just sort of grazes for the rest of the day. It's kinda lame, but it's all we've got.

Bill's birthday falls in late August. Bill is probably my favorite coworker in the history of the world. He's fiftysomething, about twenty years my senior, and has worked at the library (as a reference librarian) since the year I was born. His hair and close-cropped beard are white, his blue eyes sparkle behind his glasses, and he has an open, friendly smile. He's also a big guy—a little less than three hundred pounds—and *nobody* messes with him. He's always treating me to lunch, fetching me cans of Diet Dr Pepper from the vending machine, finding the funniest stuff online (Herbert Kornfeld's column in the *Onion*, for example, or Engrish.com, a site that documents hilarious uses of the English language in Japanese advertising), and keeping my spirits up when things get especially miserable around here. Bill is also responsible for turning me on to the silly, operatic pleasures of Godzilla movies, and our neighboring cubicles are decorated with Godzilla posters and toys, most notably a big plastic rendering of Godzilla himself. So when Bill's birthday came around, it was an honor and a pleasure to be the one to provide the cake—that is, procure the cake from the bakery section of the nearest Safeway—for his party.

My main mode of transportation is an orange and silver BMW 650 CS motorcycle with a blue weatherproof canvas bag above the rear tire. I can carry loads of things in that bag (pocketbook, water bottle, change of clothes, shoes, books, etc.), and thought that carrying a cake in it would be (heh) a piece of cake. I figured if I put the cake box in the bag, then secured it by packing my work clothes and rain gear tightly around it, the yummy German chocolate cake (Bill's favorite!) would survive the ride to work. I figured wrong.

I was working the noon–8 P.M. shift on Bill's birthday, so I'd asked a coworker to set up the traditional blue-tablecloth-covered-table in advance. I rode carefully to work and parked the motorcycle on the sidewalk in front of the library. Then I detached the bag from the back of my bike, carried it up the stairs to the office, unzipped it, removed the box, and placed it on the table in front of the HAPPY BIRTHDAY, BILL! sign.

But when I opened the lid of the box, I discovered that my transportation plan hadn't worked so well. The cake had shifted in transit, and now one side was completely smooshed against the side of the box. I carefully removed the cake from the box to see if I could sculpt it back into shape, but that made it even worse—most of the damaged icing stuck to the box, leaving a big naked patch of icingless cake. Bill's once-beautiful birthday cake looked terrible, and no amount of creative re-icing could save it. The party was due to start any minute; I had no time to get another cake. Now, Bill and the other folks in my office are not the types to turn their noses up at free food, no matter how smooshed, but I was not about to serve a *maimed cake* to my favorite colleague.

My heart sank. Tears welled in my eyes. Bill's cake! I destroyed Bill's cake!

But I didn't do it on purpose, I told myself. It wasn't my fault. It was the laws of nature. Physics or something. Yes— it was physics that caused the cake to lurch forward whenever I hit the brakes. It was physics that transformed Bill's lovely cake into something that looked like Godzilla had torn up.

AHA! Godzilla . . . of course! *If I can't fix the cake*, I thought to myself, *I'll just make it look like the damage was intentional.* I ran to my cube and retrieved my big plastic Godzilla doll.

After giving Godzilla a lovely antibacterial bath in the washroom sink and thoroughly drying him off, I planted his feet directly down into the smashed part of the cake. I put little smudges of icing on his nose and in his hands to make it look like he had helped himself to a bite. And as a final touch, I took a yellow Post-it note and fashioned it into a cone-shaped birthday hat and taped it to Godzilla's head.

The result? Bill *loved* it. His sparkly eyes sparkled even more, he started speaking excitedly in broken Japanese (yes, he knows a little), and he even helped Godzilla stomp the cake a little more. The rest of our coworkers enjoyed it, too. (Some even took pictures.) And nobody knew that it was me—not Godzilla, and not even physics—who maimed Bill's cake.

Holly Tominack lives in Baltimore, Maryland, and is married to a Johns Hopkins psychiatrist whose entire life has been a busman's holiday since the day they met.

The Scent of War

by Bonnie Lynn Dunlop

"ANOTHER DAY WITH NO WATER!" RANTED CAROL, OUR CIVILIAN interpreter. Carol was a robust, graying widow who normally wore her body armor without complaint as we trudged around the dust and debris of Abu Ghraib prison.

She stormed into our windowless quarters and flicked on the light switch. The room was a fifteen-foot square with a ceiling that was higher than the walls around it. Except for the raised plywood boards that passed for beds, there was no furniture except a steel gray Army-issue wall locker. *House Beautiful* would have called it "rugged chic."

"There hasn't been a water delivery in three days!" she moaned. "I smell so bad even the mosquitoes won't come near me!"

"The convoy must be running into trouble with insurgents along the route," I reasoned, my voice thick with sleep.

Carol was one of three females that belonged to our "mobile training team," the other two being Drew, an Army interrogator, and yours truly, the "judge advocate general," which is just a fancy way of saying Army lawyer. We'd all been sent here in March of 2005, in the aftermath of the scandal that made this place infamous. Broadly, our job was to train Iraqis in the art of (legal) interrogation. My specific role was to observe our new interrogators and make sure they didn't make the same mistakes our disgraced predecessors—it pains me to call them soldiers—had made. In my spare time, I fantasized about suing the recruiter who promised me that the Army never sent its lawyers to war.

Drew entered the room behind Carol. A petite mother of three in her early thirties, Drew hardly looks like the former drill sergeant that she is. She and Carol are both Muslim, and they'd risen at 04:45 for morning prayers.

I had been sleeping in, which means my alarm was set for 05:15. As the florescent glare of the overhead lights assaulted my eyes, I heard the familiar clank of metal as Drew laid her M16 on top of her helmet.

I propped myself up on my elbows but made no attempt to wiggle out of my sleeping bag. The smell from the bag assaulted my nostrils; it reminded me of an unwashed gym sock that had been left around a locker room for several years. The stench was a combination of the person who had been issued the bag before me—no, it hadn't been washed before being reissued—and my own odor, which had been increasing over the last three waterless days. Since I had no means of washing either myself or the bag, I decided to look on the bright side. Why get up if there's no water to bathe in? Now I could probably stay in this nasty sleeping bag till 05:30.

I reached under my "bed," grabbed my travel-size plastic box of baby wipes, and offered it to my teammates. Carol just glared at the box (she liked to wallow in misery), but Drew pulled out a couple of sheets.

"Thanks," she said.

Drew had stopped addressing me as "ma'am" after we arrived two weeks before. All three of us were now "sterile," an Army term that means we weren't supposed to communicate in a way that would reveal anyone's rank. We used aliases, too, to conceal our real identities (and I'm using the same aliases in this story).

Drew wiped under her arms and told me, "Make sure you leave some wipes for your boys—they're gonna need 'em, too."

The boys Drew was referring to were the Iraqi interrogators we were tasked to train. Their quarters were in the same building as ours, so we were together almost constantly. Initially, I had steeled myself to be distrustful of them. My commanding officer had instructed me to assume they were all dangerous spies, capable of fashioning an explosive device from my hairpins if left unsupervised for even just a few seconds. But I softened almost immediately after meeting them. One came dressed in Iraqi Special Forces regalia but spoke softly and respectfully when he described the work he'd been doing with U.S. soldiers. Another quizzed me relentlessly on American country music. (I had to explain that the ring of fire Johnny Cash sang about had nothing to do with sexually transmitted diseases.) And several of them told me heartbreaking stories of how beautiful Iraq was before Saddam, and how desperately they wanted that life again.

Unfortunately, the boys had arrived with nothing more than the clothes on their backs. Apparently no one had bothered to tell them that once the forty-five-day training program began, they wouldn't be allowed to leave the Abu Ghraib complex, for security reasons. (Again, the Army was concerned about infiltration by insurgent spies.) No going home to their families at night, no watching Iraqi soaps on television, no more Turkish coffee in the cafés—and no way to do their laundry. We had no washing machines on the post, so we sent our clothes out to be laundered. But as I mentioned, the boys had only one set of clothes. It was a real dilemma.

The simple solution would have been to buy them some more clothes. And there was, in fact, a shop on the post. Unfortunately, all they had were souvenir T-shirts that said things like HAPPINESS IS ABU GHRAIB IN MY REARVIEW MIRROR. Somehow, suggesting to Iraqi citizens that they purchase Abu Ghraib memorabilia didn't seem like a step in the right direction toward winning the hearts and minds of the Iraqi people.

In the first few days of the program, the boys, without a word of complaint, had taken to washing their clothes in the mobile sinks assembled in a small white trailer outside our quarters. But now the water supply had dried up. And even though it was only springtime, the temperature reached one hundred degrees on most days. There was only so much a box of baby wipes could do. Now the smell from the six MTT members, combined with the scent of seventeen sets of unwashed clothes, made our entire hovel smell like a decomposing desert rat that had died as a result of devouring too many raw onions.

"Why can't someone do something about the stench?" begged Carol, a look of genuine agony on her face.

It was a rhetorical question. But right at that moment, my eyes fell on a glass bottle of Tommy Hilfiger men's cologne that was sitting on a shelf in the wall locker. My boyfriend,

Sean, had given it to me prior to my deployment. "Whenever you start missing me," he said, "just smell the bottle and I'll be there." And I had. But now it was time to use it in a less romantic way. First I sprayed myself, then offered the bottle to Carol and Drew, who both accepted.

I considered offering some to the boys, but that just seemed rude. Then another idea began to gel in my mind. When the Iraqis had first arrived, Carol told me they were enthralled by all things American. So—partly out of a genuine wish to share my culture and partly to amuse myself—I'd taken to leaving little gifts for them on the table in the coffee area. First I left a box of Oreos. Then a bag of Snickers. Then some marshmallow Peeps leftover from an Easter care package. Each time, they would devour my gifts in no time flat.

"Carol," I said, "your suffering may soon come to an end."

I rushed into the coffee area and placed the bottle of cologne right next to a box of (now empty) Nature Valley Granola Bars. And within a few hours, the stench had been replaced—or at least masked—by the scent of Sean.

Bonnie Lynn Dunlop is a judge advocate in the U.S. Army. She is currently stationed in Marheim, Germany, with her only child, a boxer named Natasha.

The Accidental Potluck Party

by Michelle Cromer

"Let's have a party!" was the first thing I said when Barry, my husband, announced he was being considered as the chief of his orthopedic surgery group.

Being a good Texas girl, I thought it would be a great idea to host a luncheon at our house for all the wives. I am a partner at an advertising agency, an author, and the only wife out of the twenty-seven who works, so I'd never really had the opportunity to get to know these gals. I didn't really fit into their herd. It's a high-maintenance group, on the princess side. But any excuse to throw a party, and hopefully this effort would help my husband. Barry rolled his eyes a little at the idea—the luncheon wouldn't be cheap—but he also muttered a faint "thank you." He understood that it was smart politics.

I don't cook—or should I say, my family has asked me not to cook. The last time I tried, I put a plastic meat thermometer into a pork loin and baked it. So I hire a caterer for the food. I always take care of the other details—the overall theme, the invitations, the flowers and decorations, mixing the drinks, etc.—but leave the food to the professionals.

The day of the party, the weather cooperated magnificently. My backyard had never looked better. The water in the pool was a gorgeous turquoise, the purple gladiolas and the yellow sunflowers seemed to bloom on cue, even the birds were singing. The uniformed servers I'd hired were in place; the mojitos were chilling in crystal pitchers with freshly cut limes.

The first guest arrived at high noon, and the other twenty-five ladies arrived within ten minutes of one another. But, uncharacteristically, my caterer was late. I wasn't worried. We'd had three phone conversations in the past two weeks, and she had always been reliable. But by 12:15 I was getting concerned. And at 12:30, I reached for my cell phone.

"Where are you?" I demanded when she answered. "Everybody's here, and we have no food."

Silence.

"Michelle . . . OH MY GOD . . . I AM MORTIFIED. . . . I FORGOT," said the new sister of Satan.

I had no time for my usual slice-and-dice routine, although I do remember saying something about cutting her head off and serving it on a platter. I hung up, feeling the panic wash over me. I had no Plan B—no backup, nothing. I was standing in the middle of the nine circles of hell. I tore open all of my cabinets. I prayed for a miracle, or at least some bean dip. I took a quick inventory. A can of green beans, a box of macaroni and cheese, a can of ravioli with meatballs. Could I serve macaroni? Would they notice? Can I make that?

I picked up the phone and called my country club. I had been a board member for four years and had never cashed in my favor chip. The head chef answered.

"Please, for the love of God," I pleaded/screeched, "bring anything you have in the refrigerator, take things off people's plates, I don't care, just bring me some food, fast."

"No problem, Mrs. Cromer," he replied with confidence. "We will be there in less than an hour."

Thinking I had solved the problem, I walked across my kitchen and looked through the window. My backyard looked like the second floor of Bergdorf Goodman. All the big boys were out there: Ralph Lauren, Giorgio Armani, Michael Kors. One size zero even wore Valentino. And then I saw the tall, blonde, runway-model-looking one sneak a peek at her watch. The dark, petite one pointed at the empty plates in front of them. An hour? I had inside of fifteen minutes before these babes bolted. And what would they tell their husbands, each of whom had a vote in Barry's future? This was not good.

Just as I was about to fake a heart attack (anything to get out of here), I turned and looked out my kitchen window, and the sun seemed to shine directly onto Mrs. Elizondo's white stucco house across the street. "That's it," I said under my breath, "I will go door-to-door and beg for food."

The only problem with this solution was that I'd never met my neighbors—not even one. We live in one of the oldest sections of El Paso, and most of my neighbors are a day away from a toe tag. Mixed-up mail is the only reason I knew any of their names, and I'd lived there for five years. "I hope to hell they have something other than green Jell-O," I muttered to myself as I sprinted out my back door and across the street.

In less than ten minutes (in my new chartreuse Manolo Blahniks, no less) I had gone to six houses, introduced myself, pointed to my house, explained I was their neighbor, and asked if they had any food for the stylish yet angry mob in my backyard. To my astonishment, no one threatened to call the cops. Mrs. Elizondo, who most likely had not had a visitor since the disco era, was downright chatty. She spoke in her perfect Spanish as I explained my situation in my not-so-perfect Spanglish. When she tried to hand me her cat, I knew we had a breakdown in communication. I smiled as I shook my head and walked across her living room, into her clean kitchen and over to her refrigerator. And there I hit paydirt: an astonishing bounty of homemade Tex-Mex. My other neighbors were nearly as neighborly: my scavenger hunt scored salsa and chips, chile con queso, fresh jalapeños, nuts, and, yes, pigs in a blanket. I felt like I'd stepped into a Southern revival. I was saved!

In the end, the party was a complete success. The country club did eventually deliver, but by then I already had so much food, I packed the club food into brown bags and sent everyone home with a snack for their husbands. Several of the ladies asked me for the chili con queso recipe, which I provided after Mrs. Elizondo translated it for me. A few of the wives became my friends, I finally met my neighbors, and, yes, my husband got elected.

Michelle Cromer owns and operates her own advertising agency in El Paso, Texas. She now has a new caterer.

Stilettos in the Slush

by Marie Coolman

SOME WOMEN CAN WALK IN HEELS—AND I AM ONE OF THOSE women. I always laughed at those silly girls who carried their pumps to work in purses and totes, ruining an otherwise perfectly coordinated outfit with running shoes. But that was before I moved from L.A. to New York City. Decked out in my most stylish work clothes for my new PR job, my first few days as a commuter were spent dancing around the sidewalk grates that seemed to reach out and snap at my heels like little steel Chihuahuas. Maybe those silly girls were on to something.

This gave me something new to worry about. Okay, I could bear wearing sneakers with my fancy pants—who would notice another yuppie chick in professional slacks and sporty shoes? But the people who make the pants haven't yet figured out how to let you have one hemline for the street and another for the suite. I tried folding them up in a cuff, but the nicer the fabric, it seemed, the more slippery it was—and the greater the likelihood that I would arrive at work with nasty city sludge on my pants.

I first realized this on a chilly January morning, when the previous week's snow had turned to slushy brown puddles on the sidewalks. I had just left Grand Central Station for a meeting with one of our most important clients, a woman known for sizing people up in fifteen seconds and never changing her opinion. Headlines are my business, and with one false step, I could easily envision the announcement of my new job turning into an obituary for my career. I absolutely could not arrive looking like I'd tromped through a sewer.

Under my left arm I was carrying *The New York Times*. If only I hadn't tossed out the plastic bag it had come in, I could have used it as a sock and kept at least one leg dry. Looking around, I saw no cabs, no rickshaws, no volunteers to carry me on their shoulders. It was trudge along or be late for the meeting—and the only thing worse than arriving messy was arriving late. As panic set in, I tiptoed my way to a dry corner of the sidewalk and frantically began rummaging through my briefcase, the contents of which have saved me more than once. When I was pregnant and couldn't button my pants (why are all my problems related to pants?), I'd found a nice thick rubber band, threaded it through the buttonhole and wrapped both ends around the button to make an instant waist extender.

My briefcase didn't fail me this time, either: Attached to a packet of press releases and pitch letters, I found four large, silver paper clips. Without wasting any time, I paper-clipped a temporary hem next to both the inseam and the outseam, keeping the fabric safe from the dirty film covering the street. Giddy with relief, I kicked up my feet, half to test the clips and half with an "if you can make it here, you can make it anywhere" dance step.

As I made it safely and cleanly to the office, I wondered how many of the hundreds of people I'd walked by that morning had silently mocked me (as I would have if I hadn't been me). And then I realized that I didn't care if anyone else thought I was ridiculous. The rubber-band trick kept me in my favorite pants months longer than a pregnant lady should have been able to pull off. And my new paper-clip trick was going to be my daily solution to the hem dilemma.

Marie Coolman works in public relations in New York. Her other million-dollar idea—an adjustable tube of sunscreen that offers a variety of SPF options—has not come to fruition, and she has the freckles to prove it.

Where in the World Is Ricky Williams?

by Chris Jones

WHEN RUNNING BACK RICKY WILLIAMS DECIDED TO QUIT PRO football and travel around the world in 2004—leaving bewildered teammates, angry fans, and a thousand unanswered questions in his wake—I decided that I wanted to write about him for *Esquire*. Having spent my entire career writing about sports for a living, I was desperate to cover something that didn't take place on a field or in an arena—to stretch my legs a little. Ricky seemed like the perfect subject to chase. The only problem was, I had no idea where to find him. No one did. Over the past few months, the former all-pro running back for the Miami Dolphins had been on an epic, meandering journey that had taken him to Jamaica, the Bahamas, England, across Europe (following Lenny Kravitz on tour), Samoa, Fiji, Hawaii, and Japan. Now, who knew?

After rustling up his e-mail address—how I did that I'll need to keep secret—and explaining to Ricky that I really, really wanted to tell the story of his running away, I took the plunge and asked him where he was hanging out. "Australia," he wrote back. "Near the ocean. In a tent."

And that was it.

He did promise that if I found him, he would be welcoming and honest. But he made it plain that he wasn't going to help me out any more than he already had. It felt like a test: Before I could write about his journey, I would have to take my own.

I was happy for the adventure. However, my boss wanted some assurance that whatever money he spent to fly me to the other side of the world wouldn't go to waste. He knew I couldn't guarantee him that I'd find Ricky, but he wanted me to give him a reasonable chance of success. I began to calculate the odds: a man, in a tent, in the only country big enough to double as a continent. Less than 1 percent, I decided. I told my boss fifty-fifty.

That, it turns out, was good enough. Within a few days, I would be flying to Australia. But where in Australia, exactly? My first trick was narrowing my scope a little. Fortunately, when I was younger, I had lived Down Under for awhile, and that made things easier. I could eliminate those places I knew he wasn't. If he was in a tent, he wasn't anywhere near the Sydney Opera House. And if he was near the ocean, he couldn't be lost in the Outback. Then I looked at the places I had left. There were still quite a few, but one jumped out. There is a small town called Byron Bay on Australia's east coast between Sydney and Brisbane. I remembered it as a surfer's paradise, with a long white-sand beach, quiet except for the hippies playing their bongos. It was exactly the sort of place I could see Ricky taking to. At the very least, it seemed like a good place to start.

I flew to Brisbane, had a free shower in the airport (every airport could take a lesson from Brisbane's opulent cans), rented a white Honda hatchback, and began heading south. It's about a two-hour drive, but it felt more like four days. My jet-lagged mind was consumed with what might happen next. Even if Ricky was in Byron Bay, how would I find him? If he wasn't, where would I go next? *Maybe this was all a big mistake*, I thought, and I started to feel hot with regret. For all I knew, Ricky had picked up and gone trekking in the mountains of Nepal, and here I was, just pulling into Byron Bay. The streets were crowded; the shops and restaurants were busy; it looked as though a hundred surfers were bobbing in the water. Was Ricky one of them?

I parked, turned off the engine, and stared out the window at the passing crowds. Now what? In that moment, filled with panic and dread—but also with excitement and a strange kind of wonder, shaking my head at the mess I'd gotten myself into—I remembered that sometimes the best solution to a problem is the simplest. Ricky was in a tent, that much I knew. According to a map I'd picked up, Byron Bay had three campgrounds. I would go to each one and ask if anyone had seen a big, strong, African American named Ricky.

He wasn't at the first place.

He wasn't at the second place.

But at the third, an older man with leather for skin pointed across town toward something called the Arts Factory, an out-of-the-way campground on the edge of the swamps. It hadn't been on my map. I hadn't known that it existed. "I bet he's there," the man said.

I followed his directions, walking down a long, winding road out of town. And there, nestled among the tea trees, rose the Arts Factory. A woman sat behind a desk in a great thatched hut, and I asked her if a big, strong, African American named Ricky was staying there. "Yes," she said. "He's in one of the tents."

I just about exploded on the spot.

That is, until I saw there were more than a hundred tents. I was just about to start poking around, one by one, when I heard a familiar voice coming out of a small hut nearby. It smelled like a kitchen. I tiptoed over and peeked through the door. And there was the object of my obsession, squeezing the juice out of a giant stalk of celery.

"Ricky?" I said.

He turned around and smiled.

A man, in a tent, in the only country big enough to double as a continent. And there he was—and what turned out to be the sports story of my career—just like that.

Chris Jones, now living in Ottawa, Canada, is a practicing fatalist who isn't responsible for his actions, especially if you get between him and a taco. His story on Ricky Williams— titled "The Runaway"—was called "the most entertaining piece of the year" in 2005 by ESPN.com.

The Day I Managed to Untie Myself from a Rocket Loaded with TNT and Save the Presidential Palace of El Santori in the Nick of Time

by Stacey Grenrock Woods

I RECALL BEING SOMEWHERE ONCE AND SOMEONE DID SOMETHING like put some chewed gum on the end of an unfolded paper clip and used it to fish a dollar from behind the refrigerator, and my friend, Siobhan, said, "Nice work, MacGyver." I didn't completely get it, since all I know about *MacGyver* was that it starred a three-named actor who had a sort of hedgehog burst of hair and who was not Jan-Michael Vincent.

The preceding sums up the full extent of my familiarity with MacGyver—the show, and the reference—so you can understand why it was with some trepidation that I accepted Brendan Vaughan's invitation to contribute to this collection.

He told me that, of course, I didn't have to submit anything, and then he made a joke about how it'd be okay—he'd get over it "in about eight months." I know most of you are thinking, "Oh, what a funny guy! He is not serious." Think again.

You see, as my editor at *Esquire* magazine, Brendan has the power to make my life just terrible. He doesn't—the editing of my monthly sex column is generally quite painless—but the point is, he *could*. If provoked, Vaughan could, on a whim, devastate line upon line of my precisely chiseled prose with no more than a "Is there something funnier you can say here?"

I thought and thought and thought. When in my life have I ever gotten out of a tight spot with cunning and grace? When have I ever thought fast and acted faster? The answer, friends, is "never," and this is why I always seem to end up at some utter stranger's baby shower, playing "Guess What's in the Poopy Diaper?" with a group of pockmarked menthol smokers.

Certainly I've gotten out of things, just not well. Leaving phone messages when I know no one will be around? That's one of mine. And the hangover and the well-placed public trash can? That was just dumb luck.

It's a sad day when you realize you are not a smooth operator. I barely even have any colorful stories to tell. There is that one—it's more of a cautionary tale, really—about the time I watched helplessly as my parakeet blasted like a rocket out of his cage and disappeared into the sky. By the time I get to the moral ("Never clean your bird's cage during an eclipse") my listeners aren't exactly enthralled.

As the weeks progressed, Brendan needed to make sure I was still on board. More e-mails volleyed back and forth containing various reassurances (from both sides). He even came up with a motivational deadline: "October 17." He just wasn't going to let this thing go. But October 17 came and went like a lamb, like we both knew it would.

It was becoming more and more apparent that the tight spot I needed to cleverly get out of was this.

Yes. The irony was as sharp as the end of a paper clip. The subtlety as soft as some chewed gum one might put on the end of that paper clip to pick up more irony from behind an allegorical refrigerator.

I could certainly get five hundred words out of this sad tale here—the one about weaseling out of the commitment—but I like to think I'm better than that. Besides, I can just hear Vaughan: "Very funny, Stacey, very Charlie Kaufman–esque, very meta, very '97. I can't use it."

So I did the only thing I could do: I took a handful of Ambien and sat on the couch.

And then, the most amazing thing happened! I suddenly remembered this absolutely fantastic story from my past:

So there I was, being tied to a rocket that was loaded with a ton of TNT. I had only five minutes to get myself and a hostage (who looked eerily like the actress Kay Lenz) untied and stop the blast from blowing up the presidential palace of

El Santori. Luckily, I spied a cart of liquid nitrogen tanks nearby. I threw my shoe at it, knocking the release lever free, and the cart rolled toward me. But then, a crack in the pavement stopped it. I took off my other shoe and looped the laces around the cart's handle. The cart rolled forward into the launch pad, causing one of the tanks of liquid-nitrogen to fall just into my reach. I activated it, freezing the ropes until they broke and, with seconds to spare, I freed myself and Kay Lenz, stuffed a tarp into the rocket's air intake valve, and fled to safety.

Whew! What a tale! Sure, I lifted most all of it from a *MacGyver* plot I found on the Internet (www.rusted-crush.com/macgvyer/macep128.html, Episode #128: "Obsessed"; air date: September 30, 1991), but did it get me out of a jam? For now. Is it clever? Oh, yes. Is it too clever? Perhaps. Time (and Vaughan) will tell.

I believe it was Truman Capote who said, "Too clever is dumb." Those are strong words for a man who has acted alongside James Coco, but lucky for me, there has been a very clever new invention since the time of Angus MacGyver. It is known as "caller ID," and when Brendan calls, I intend to use it.

Stacey Grenrock Woods is a "writer, actress, whatever" (WAW) who lives in Los Angeles. You can read about all the people who have let her down in her upcoming memoir, I, California (Scribner) due out sometime real soon.

Acknowledgments

I'M INDEBTED TO COUNTLESS PEOPLE WHO HELPED MAKE THIS book, but none more than my editor at Hudson Street Press, Danielle Friedman, who came up with the idea in the first place (an especially impressive feat since she was four years old when *MacGyver* debuted). The book would not exist without Danielle's care, creativity, and dogged pursuit of clarity, especially in the chapter about cars.

I want to thank a few other Hudson Streeters: Laureen Rowland, under whose authority this book was published; Liz Keenan, for promoting the hell out of it; Lucy Kim, for designing the jacket; Eve Kirch, for designing the interior; and Abigail Powers, Norina Frabotta, Susan Schwartz, and Fabiana Leme for keeping me on schedule.

I am enormously grateful to every single person who visited WhatWouldMacGyverDo.com, but especially to everyone who submitted a story, and even more especially to those whose stories were selected and who patiently revised and revised and revised until we ran out of time and just couldn't revise anymore.

Speaking of WhatWouldMacGyverDo.com, thanks to Josh Mack for building and maintaining the site.

An unknowable number of people helped spread the word about this project, and if I knew who they were, I'd thank them all. But I don't, and that's the beauty of the Internet. However, I would like to specifically thank Shoshana Berger and Katharine Sharpe at *ReadyMade* magazine; Robin Hemley at the University of Iowa; Dana Miller at the Gotham Writers' Workshop, and all the bloggers and *MacGyver* fan-site editors who mentioned the book, especially Rockatteer at MacGyverOnline.org.

Thanks to my wise agent, Sloan Harris, and his kick-ass assistant, Katharine Cluverius.

Thanks to my boss at *Esquire*, the great David Granger, who permitted me to do this on the side.

Finally, I'd like to thank my beloved wife, Melissa, and son, Roan, who spent many a Saturday without me as I toiled away on the manuscript. Melissa, I wish I could say this project made me a better handyman. I can't. But now that it's over, at least I'll have more time to practice.

Index